Radio Frequency System Architecture and Design

For a listing of recent titles in the
Microwave Library,
turn to the back of this book.

Radio Frequency System Architecture and Design

John W. M. Rogers
Calvin Plett
Ian Marsland

**ARTECH
HOUSE**

BOSTON | LONDON
artechhouse.com

Library of Congress Cataloging-in-Publication Data
A catalog record for this book is available from the U.S. Library of Congress.

British Library Cataloguing in Publication Data
A catalog record for this book is available from the British Library.

ISBN-13: 978-1-60807-537-9

Cover design by Vicki Kane

John Rogers dedicates this book to his daughter, Serena.

Ian Marsland dedicates this book to his wife, Kelly Harrison, and their children, Kathryn and Bridget.

Contents

Preface

This work started as a set of notes for a graduate course. Having a background in RFIC design, it became apparent that many students, even after completing Ph.D. degrees in radio hardware, had very little knowledge of systems-level specifications for the circuits that they were becoming experts in designing. In contrast, colleagues working at the systems level in communications seemed to be very concerned with their own issues and challenges, leaving them with little time to consider the non-idealities of the radio. After much reflection, it seemed that there was a hole in the knowledge that we were providing to our hardware designers. This book is an attempt to fill that hole and provide some insight into the design process for a radio front end at a higher level. In this book, we will not talk about the intricacies of circuit design at the transistor level, but instead take a step back and look at a radio from a higher level in order to understand how all the circuits operate together to provide an overall working system.

Writing this book has been a learning experience as much for us as it may be for many who read it. As always, we hope that what you read helps you with your own journey towards more understanding in this field. In spite of our best intentions, errors will likely be found in this work. We will maintain a list of errors on a web page (www.doe.carleton.ca/~cp/systems_errata.html) that readers may access. Please let us know if you find additional errors not yet listed on this page.

Three professors contributed to this work. We have all had long careers that have been touched by colleagues, former teachers, and students who have all helped us with our understanding and professional growth. As the list is very long, we have not included it here, but we would like to thank all of you who have helped us with this journey.

CHAPTER 1

Introduction to RF Systems Design

1.1 Introduction

Wireless technology is now ubiquitous. Over the past two decades, so many communication devices have become part of our daily lives that it boggles the mind to think about them all. Now people connect their computers wirelessly to the Internet using wireless local area networks in the office and home as well as public places like coffee shops. We carry cell phones that provide not only voice, but also text and Internet access. In our cars we have GPS for navigation and Bluetooth transceivers for wireless headsets to allow us to talk while driving. These are just a few of the commercial devices in our everyday lives that make use of wireless technology.

As our communication equipment becomes more sophisticated, so do the radios and other hardware required to enable that communication. In the past, a communication device could be designed by only a handful of people, but now hundreds or maybe even thousands are involved at different levels of the design. Many specialists are required at many different levels to make a product and perhaps now no one single person could ever have all the required skills to know or design a whole system. As these devices become ever more complicated, it becomes harder for us as engineers to understand how the bits and pieces on which we work fit into the grander scheme of things. It is the goal of this book to help the readers to expand their vision beyond the scope of their own detailed research and help them to see how all the separate parts fit together to make a whole device. This book will focus on the skills necessary to design a radio at the block level. To this end, common radio architectures will be considered and the reason why blocks are connected the way they are will be studied. As well, detailed calculations and the theory to determine block-level specifications will be discussed. Because many of these calculations require an understanding of the workings of the back end of the radio, basic theory and operational concepts related to this part of a communication device will also be considered.

1.2 What is a Radio and Why Do We Need One?

We all use radios in our everyday lives and many students and engineers have worked on them, but how often do we really stop and think about exactly what a radio is and why we need one? Let us start with a very simple definition of a radio: A radio is a communication device that allows information to travel wirelessly from one place to another over some medium. The medium most often takes the form of air, but electromagnetic signals also propagate to some degree through walls,

1

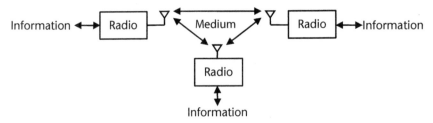

Figure 1.1   A conceptual drawing of a radio's function.

trees, water, and so forth. An example of three radios sharing a medium is shown in Figure 1.1.

The previous definition of the radio can be expanded to make it more technically precise. A radio is a communication device that takes some information, converts that information into the form of an electrical signal, takes the electrical signal, and transmits it across the medium using only a predetermined band of frequencies. It also reverses this process for signals that it receives over the medium from other radios.

Having defined the radio, the next question to consider is why do we need one at all? Speaking and listening can be seen as a form of wireless communication. We all speak and listen using signals (voice) at (approximately) the same frequencies, and this is why we are able to communicate with each other successfully most of the time. Because our voices only carry for a relatively short distance, we do not interfere with conversations in other rooms, buildings, or cities. Even when a group of people are all in one room, although we hear everything, we are able to focus on one person's voice and filter out the rest of the "noise." Our biology and brains are truly amazing and, with the radios we have today, we are unable to duplicate much of this performance. If a number of radios all "talked" at the same frequencies, no one would be able to "hear" anything. This is why the signal from each different radio makes use of a different part of the frequency spectrum and so interference is avoided. This ability of the radio to translate each conversation into a different part of the spectrum is an important reason why we need a radio. A very important consideration for radios when transmitting over the air is that the size of an antenna is inversely proportional to the frequency at which it must operate. Thus, higher frequencies mean smaller antennas, and if the aim is to have small wireless devices, then the frequency they transmit and receive at must be high.

Most of this book will deal with the design of transceivers that use wireless links and are therefore radios; however, it should be noted that while many applications for the theory presented in this work are wireless, there are some notable exceptions. For example, cable tuners contain receivers, but in this case the signals are carried from source to destination by coaxial cables and no antenna is required.

1.3   The Radio Spectrum

Because anyone transmitting over the air has the potential to affect everyone, this is an area in which the government in every country gets involved in the engineering. It is up to the laws in each country to regulate who can use what part of the

Table 1.1 Some Well-Known Frequency Allocations for Different Applications

Frequency Band	Use
535 kHz–1,605 kHz	AM Radio
54 MHz–72 MHz	TV Channels 2–4
76 MHz–88 MHz	TV Channels 5 and 6
88 MHz–108 MHz	FM Radio
174 MHz–216 MHz	TV Channels 7–13
512 MHz–608 MHz	TV Channels 21–36
614 MHz–698 MHz	TV
824 MHz–849 MHz Transmit, 869 MHz–894 MHz Receive	Cellular
902 MHz–928 MHz	ISM and cordless phones
1.57542 GHz	GPS
1,710 MHz–1,785 MHz Transmit, 1,805 MHz–1,880 MHz Receive	Cellular
1,850 MHz–1,910 MHz Transmit, 1,930 MHz–1,990 MHz Receive	Cellular
2.4 GHz–2.5 GHz	ISM, cordless phones, and WLAN
5.15 GHz–5.35 GHz	WLAN
3.168 GHz–10.552 GHz	Ultrawideband (UWB)

radio spectrum and for what purpose. For example, the military will have dedicated parts of the spectrum reserved for the use of their radar systems, and companies with cellular networks have paid very large amounts of money to use particular bands exclusively for their networks. Other bands are designated as "unlicensed," meaning that they are free for anyone to use provided that they obey the rules of transmitted power. The industrial, scientific, and medical (ISM) bands are examples of these unlicensed bands. Some common uses of different frequency bands are given in Table 1.1, including the frequency allocations for AM and FM radio, TV channels that are broadcast over the air, and many of the different cellular bands. Two of the ISM bands and their common uses, such as cordless phones and wireless local area networks (WLAN), are included as well. UWB is also listed in the table and operates at a low power over a very wide range of frequencies (even across frequencies allocated to other applications). It is allowed to do this as long as the transmit power is very low.

1.4 A Communication Device

Engineers are often very focused on the one part of a communication device that is related to their area of expertise, and may have only a vague understanding of how the overall system works. A block diagram of a typical wireless communication device is shown in Figure 1.2. In a modern communication device the actual hardware and software that is involved in communicating is usually only a small part of the overall device. At the heart is the applications processor, which runs the device's operating system. In addition to running user applications such as playing music or

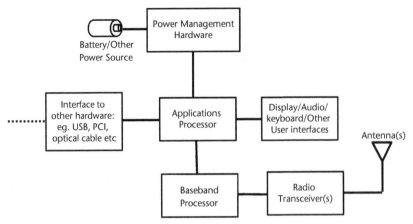

Figure 1.2 A very high-level block diagram of a communication device.

games, the operating systems controls all the hardware components of the device. The device will likely contain a variety of user interface elements such as a speaker and microphone, keyboard or touch-sensitive screen, a display, and a camera (or two). There will also be hardware and software involved in managing the power source of the device. This circuitry may clean up the power supply noise, create other voltage references, and monitor if the voltage gets too low or too high. The device may interface with other hardware via a host of interfaces such as USB, PCI or other, particularly if the device is embedded into a computer (such as a WLAN card). There will also likely be several different wireless communications interfaces, such as IEEE 802.11 WLAN, Bluetooth, and different cellular network interfaces.

When the applications processor needs to transmit some data over a wireless interface, it forwards a packet of bits to the digital baseband processor (BBP), a dedicated hardware unit that is responsible for controlling the interface. It is the job of the BBP to encode the data that the radio is transmitting and interpret the data that the radio is receiving. The BBP will also control the radio by adjusting its gain and operating frequency, changing its operating modes and turning on or off different functions of the radio. The BBP may also decide what modulation scheme to use (a higher data rate modulation if the signal is strong or a lower rate if the signal is weak). Other functions of the BBP include determining when to transmit, when to receive data, how to handle handoff between cells, and other high-level issues with communicating in an environment with many other users. The BBP performs various baseband digital signal processing operations on the transmitted data before passing it to the radio transceiver, which generates a radio-frequency analog signal suitable for transmission over the air via the attached antenna.

Building a communication device like a cell phone would likely involve the design efforts of thousands of engineers in a variety of disciplines. This book will deal only with the transceiver and some of the baseband digital signal processing. A more detailed, but still very high-level view of these two sections of the communication device is shown in Figure 1.3. The baseband signal processor starts by receiving information from the applications processor and performs a number of operations on the information. First, the data may be compressed to reduce the number of bits that must actually be transmitted. The data may then be encrypted to increase security. Channel encoding (error control coding) is then performed, whereby carefully

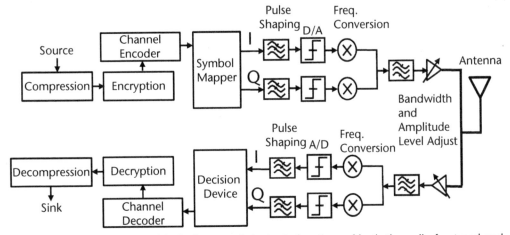

Figure 1.3 A very simple block diagram of the basic functions of both the radio front end and baseband signal processor.

controlled redundancy is added to the data to facilitate error detection and correction at the receiver. This may be as simple as adding parity bits from a cyclic redundancy check (CRC) code for error detection or from some more complicated error correcting code such as a convolutional code or a low-density parity check (LDPC) code. After this task is complete, the next step is to change the bits into baseband waveforms. The data is converted into two square waves called the in-phase (I) and quadrature phase (Q) baseband waveforms. These waveforms contain the information about the amplitude and phase of the actual transmitted waveform. The block will determine the actual amplitudes of these two waveforms depending on the type of modulation scheme to be used to transmit the data. Note that even though we now have "waveforms" rather than bits, these waveforms are likely just stored digitally as discrete-time samples and are not actual voltages and currents. After this conversion is complete, the waveforms are filtered by a digital pulse shaping filter to optimize their bandwidth (real circuits and channels cannot accommodate perfect square waves, which use infinite bandwidth). Once the pulses have been shaped, they are passed to the transceiver by way of a digital-to-analog converter. The radio front end then upconverts the signal by modulating a carrier wave, filters it to make sure there is no energy outside the desired frequency band, and amplifies the carrier to a specific power level to drive the transmitting antenna.

Once the energy has traveled across the channel, the radio front end must detect the received signal, amplify it, and return it to baseband, providing an output signal with fixed average power. Filtering at the receiver prevents the radio from having to process unwanted signals that may be present at other frequencies. Once downconverted to baseband, an analog-to-digital converter makes a digital copy of the signal for further processing. In the digital domain some further filtering (with a filter matched to the transmitted pulse shape) may be performed to maximize the signal-to-noise ratio (SNR), after which it is converted back into bits by the decision device. Once in this form, error correction is performed and the information is decrypted and decompressed. Note that some more modern channel decoders perform error correction directly on the received samples at the output of the matched filter, because this tends to decrease the likelihood of errors.

1.5   Baseband Signal Processing Versus RFIC Design

The baseband signal processing core of a communication device is responsible for collecting binary data from the applications processor and preparing it for transmission. It is also responsible for collecting the signal samples from the receiver front end and recovering the binary data for delivery to the applications processor. Whereas the BBP operates on bits and discrete-time samples, the radio transceiver operates on analog voltage waveforms, and the two components are joined by digital-to-analog and analog-to-digital converters.

Thus, the engineers who work on the baseband signal processing do so using digital circuits and their work, while challenging, does not require them to have knowledge of voltage, current, and power signals as an RFIC designer must. RFIC designers must implement circuits at the transistor level and deal with all the imperfections of the real physical world. Because signals in the BBP are represented digitally, the language used by BBP designers to describe the quality of a waveform is different than that used by analog RFIC designers, leading to a chasm of ideologies that separate these two design groups. While the BBP designer wants to talk about the bit error rates (BER) and energy per bit or symbol, the RFIC designer wants to know what noise figure the radio needs to have. The truth is that a particular noise figure is necessary for the BBP to achieve a particular BER. The problem is that few designers on either side of this design effort know how to speak the other's language. The BBP designer lives in a theoretical world in which signals are mathematical things that can be operated on. They can code up a piece of software to perform any given function and the function can be carried out in an almost ideal manner. The RFIC designer must deal with the real physical world, and real physical devices with all their imperfections. While the BBP part of the radio is often given very detailed consideration, the RF transceiver is left as an idealized mathematical model by the BBP engineer. Likewise, the RFIC designer may focus on block level design with little if any consideration of what must happen on the other side of the ADC. Thus, both sides tend to ignore the other part of the radio, in part because the block is not in their area of specialization, and in part because

Table 1.2   Relationships Between BBP and RF Design

	BBP	RF
Signals	Represented as bits or quantized discrete-time samples.	Represented as voltages, currents, or power signals.
Impementing a signal processing function	Function is implemented using software. Implementation can be ideal save for using a finite number of quantization levels.	Function is implemented at the transistor level using nonideal components.
Typical design tools	Capable of simulating complex baseband waveforms without a carrier.	Capable of simulating carriers without modulation data.
Complexity of operations	Can be very complex at the expense of hardware and power.	Implements simple functions at very high speeds with relatively low power consumption compared to a digital implementation.

simulation tools are generally poorly suited for crossing this boundary. Classic BBP simulation tools are unable to simulate RF signals with high frequency carriers, and classic RFIC design tools are too slow to meet the needs of BBP engineers. Table 1.2 attempts to illustrate some of the differences in the design approaches between these two groups. It is a major goal of this book to provide a bridge between these two groups to provide better understanding of what constraints the other works under. It is hoped that with better mutual understanding will come better radio designs.

1.6   Overview

There are many excellent textbooks on radio frequency integrated circuit (RFIC) design [1–3] and also many excellent textbooks dealing with BBP issues [4–7]. These areas have now been well documented in the literature and many universities offer courses in these areas. However, at the start of their careers, RFIC designers are usually involved in the detailed design of individual blocks. Designers are very concerned with achieving the required performance from their block and often the performance level required of the block is provided by someone else. Often the designer does not know where this specification came from or how a particular level of required performance was calculated. It is the intent of this book to attempt to present the principles used in determining block level performance requirements for a radio, which is the job of the radio architect.

For BBP engineers, this book will attempt to provide enlightenment about constraints faced by RFIC designers. It will attempt to make clear what impact nonideal RF performance will have on the quality of the signals provided to the back end. For RFIC designers, this book will attempt to make clear why the specifications and performance levels demanded of their blocks are so important to the overall radio performance and how they contribute to the quality of the communication link. The first aim of this book will be to teach enough RF-related systems knowledge and information theory to be able to translate a BBP architect's requirements into radio specifications (ideally, this task would be performed by a BBP architect working closely with an RF system architect). The second aim of this book will be that once the overall systems specifications for a radio are known, it will be possible to determine what blocks are required in the radio (a job of the RF systems architect). The third aim of this book will be to provide instruction on how to determine the required performance level for each of the blocks in the radio (also job of the RF systems architect). Therefore, readers will have a better understanding of a radio architect's tasks and skills necessary to ensure the radio design is successful. To meet these goals, the book is laid out as follows: Chapter 2 will review basic concepts related to BBP, including many basic performance metrics; Chapter 3 will review basic concepts and performance metrics related to RFIC design; Chapter 4 will discuss radio architectures; Chapter 5 will discuss how to calculate performance requirements for all the RFIC blocks and RF performance levels required by the radio; Chapter 6 will discuss frequency synthesizers and how they impact the radio and subsystems; and Chapter 7 will conclude the book with many large-scale examples of how a radio will be architected.

References

[1]   Lee, T. H., *The Design of CMOS Radio Frequency Integrated Circuits*, 2nd ed., Cambridge, U.K.: Cambridge University Press, 2003.

[2]   Razavi, B., *RF Microelectronics*, Upper Saddle River, NJ: Prentice Hall, 1998.

[3]   Rogers, J., and C. Plett, *Radio Frequency Integrated Circuit Design*, 2nd ed., Norwood, MA: Artech House, 2010.

[4]   Stremler, F. G., *Introduction to Communication Systems*, 3rd ed., Reading, MA: Addison-Wesley Publishing, 1990.

[5]   Rappaport, T. S., *Wireless Communications*, Upper Saddle River, NJ: Prentice-Hall, 1996.

[6]   Proakis, J. G., *Digital Communications*, 4th ed., New York: McGraw-Hill, 2001.

[7]   Sklar, B., *Digital Communications: Fundamentals and Applications*, 2nd ed., Upper Saddle River, NJ: Prentice Hall, 2001.

CHAPTER 2

An Introduction to Communication Systems

When we use communication devices we generally do not pay much attention to how they work. We know that if we type in the URL of a Web site into our Web browser, the desired Web page will appear a moment later. Where that content came from and how it was delivered to our device are far from our primary concern. We are more interested in how we interact with our device than how it interacts with anything else. That is, we only have a vague awareness that there is communication link between our device and somewhere else. However, the application programmers who wrote the Web browser need to be able to interpret the keystrokes (or screen taps) of the user and translate them into HTTP commands that can be transmitted to the appropriate Web site, which will respond with data packets containing an HTML document describing the requested Web page. From the perspective of an application programmer, the communication link is a simple device where bits are fed in at one end and come out the other. How this is actually achieved is not particularly relevant to the application programmer.

Although the implementation details are not important to the application programmer, there are some attributes of the link that are of interest, including the throughput and the reliability. The throughput is usually measured in terms of the number of bits that can be transmitted per second. Wireless communication links are notorious for the variability of the throughput over time as the user moves around. Reliability of a communication link usually refers to either the bit error rate (BER) or the outage probability. The BER is the probability that a transmitted bit is received incorrectly, and the outage probability is the probability that the communication link is unusable.

A communication system design engineer focuses on how to realize a system that is able to transmit bits over a physical communication channel. The design is typically based on idealized mathematical models of the various system components, and, by analyzing these models, the system engineer determines what tasks the components should perform, with the objective of maximizing the system throughput while minimizing the BER and outage probability. External constraints imposed by governmental regulatory bodies, device complexity and cost, and battery usage must also be considered.

Based on the high-level system design, RF system design engineers and DSP engineers decide how to implement the individual components to meet the system-level functional requirements. This design work is typically at the circuit level (for RF components) and software level (for DSP components). Because it is not always possible to meet the original design specifications, an iterative design approach is often needed.

This chapter provides an introduction to some of the system-level issues and constraints, along with a discussion on the basic techniques for system analysis and design. At the system level we are not so much concerned with how to implement each of the individual blocks but rather with what tasks they are supposed to perform. The actual implementation details of some of the blocks are discussed later in this book, and because it is often impossible to build these blocks to perfectly perform their intended tasks, the implications of these imperfections on system design are also investigated.

2.1 A Simple Digital Communication System

To introduce some of the issues that need to be addressed when designing a communication system, we will consider a simple example where we wish to transmit just a single bit. To communicate bits, we need some physical representation for zeros and ones that is agreed upon by the sender and the observer. For example, in written English (and all other languages that use Arabic numerals), we use 1 to represent a one and 0 to represent a zero. Romans used I to represent a one and did not write zero. Other examples in modern usage are shown in Figure 2.1.

For electronic communication, we represent zeros and ones with different electrical signals. For example, we could raise the voltage to a high level to represent a 1 and hold it low to represent a 0. This simple scheme, known as on-off keying, is widely used, but there are many other schemes that are also used, including polar return-to-zero and pulse position modulation. Each scheme has its advantages and disadvantages in terms of spectral efficiency, probability of detection error, ease of implementation, and other factors. However, as long as the transmitter and receiver agree on the representation (that is, they speak the same language), communication is possible.

In this example, we consider the transmission of just a single bit using on-off keying. We will raise the voltage on a line to A volts for T seconds to represent a one, and keep it at 0 volts to represent a zero. The transmitted signal, $v(t)$, is therefore either $s_0(t)$ or $s_1(t)$, as shown in Figure 2.2, depending on whether a 0 or a 1 is to be transmitted.

The signal propagates over the communication medium to the receiver. During transmission the signal is inevitably distorted, so the received signal is usually quite different from the transmitted signal. For example, the received signal may look something like that shown in Figure 2.3(a), which would typically arise when the transmitted signal is corrupted by the presence of additive noise, so the received signal is

$$r(t) = v(t) + w(t) \qquad (2.1)$$

where $w(t)$ is the additive noise.

Cyrillic Hindi Tamil Thai

Figure 2.1 Different graphical symbols used to represent the number 1.

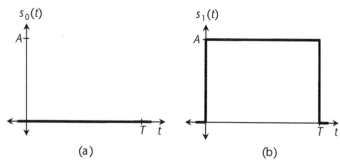

Figure 2.2 Signals for on-off keying. Signal $s_0(t)$ shown in (a) is used to represent a bit value of 0, and $s_1(t)$ shown in (b) is used to represent a 1.

The receiver, upon observing $r(t)$, must decide whether a zero or a one was transmitted. Intuitively, a quick comparison between Figure 2.2 and Figure 2.3(a) suggests that $r(t)$ looks a lot more like $s_1(t)$ than $s_0(t)$, so the receiver should decide that a one was transmitted. A more formal algorithm for the receiver to employ when making its decision could be to sample $r(t)$ at some time between 0 and T, such as at $t = T/2$, and then decide whether or not $r(T/2)$ exceeds a certain threshold. For example, if $r(T/2) \geq A/2$, then the receiver should decide that a one was transmitted, and it should decide that a zero was transmitted otherwise.

This algorithm, while simple to implement, is unfortunately not very reliable. If the signal is weaker and/or the noise is stronger, such as shown in Figure 2.3(b), it is quite likely that the receiver will make an incorrect decision, because any given sample is likely to be below the threshold, even if $s_1(t)$ was transmitted. The receiver could give better performance (that is, a lower probability of error) by better exploiting all the received information. Because the additive noise, while random, is as likely to be positive as negative and changes very quickly over time compared to the signal duration, T, it is better for the receiver to first calculate the average of the received signal,

$$R = \frac{1}{T} \int_0^T r(t) \, dt \tag{2.2}$$

This value will be close to 0 if a zero was transmitted, and close to A if a one was transmitted, so the receiver could decide that a one was transmitted if $R \geq A/2$ and choose zero otherwise.

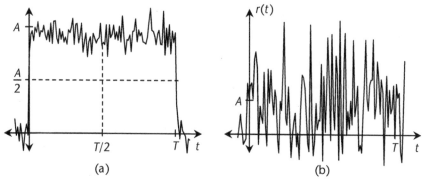

Figure 2.3 Received signal in the presence of (a) weak and (b) strong additive noise.

This simple example captures the main ideas required for system design, namely, the mapping of binary data to analog signals at the transmitter, and the recovery of the binary data from the received analog signal at the receiver. Important matters such as modulation, spectral shaping, throughput, and probability of error have been neglected. These issues will be elaborated upon in the remainder of this chapter.

2.2 Basic Modulation Schemes

Simple baseband signaling schemes such as on-off keying, while widely used in wired communication over short distances, are not suitable for wireless transmission. For example, transmission of low-frequency electromagnetic waves requires very long antennas that are not practical for portable communication devices. It is therefore necessary to convert the signal to a higher-frequency band prior to transmission using a technique known as modulation. Demodulation is used at the receiver to convert the signal back to baseband. By using a higher frequency band to transmit the signals, it is also possible to use a spectrum-sharing technique known as frequency division multiple access (FDMA), whereby many users are able to communicate simultaneously, with each user using a different frequency band.

Modulation is achieved by using the baseband data-bearing signal to modulate (or vary) the amplitude, phase, or frequency of a carrier wave, which is typically just a cosine wave. In the following sections we will discuss four common modulation techniques: amplitude shift keying, phase shift keying, frequency shift keying, and quadrature amplitude modulation.

2.2.1 Amplitude Shift Keying (ASK)

One simple modulation scheme, known as *amplitude shift keying* (ASK), involves using the amplitude of a carrier wave to indicate the value of the transmitted message symbol. A binary ASK scheme would use two distinct amplitudes to represent the value of a single message bit. For example, a sinusoid with an amplitude of 0 volts could be used to represent a bit value of 0, and a sinusoid with an amplitude of A volts could be used to represent a 1, as shown Figure 2.4. Each signal has a duration of T seconds.

With this binary ASK scheme, one bit is transmitted every T seconds. One way to transmit data more quickly is to decrease the signal duration, T. However, because the bandwidth of the transmitted signal is inversely related to T, decreasing T leads to an increase in the required bandwidth. Because the available bandwidth is typically restricted and regulated by the government, there is a limit to how small T can be. Another way to increase the data throughput is to increase the number of bits that are transmitted in each time slot. With M-ary ASK, M different signals (each with different amplitudes) are used to represent the M different values realized by $\log_2 M$ bits. The M different signals are expressed as

$$s_m(t) = \begin{cases} A_m \cos(2\pi f_c t + \phi_c) & 0 \le t \le T \\ 0 & \text{otherwise} \end{cases} \tag{2.3}$$

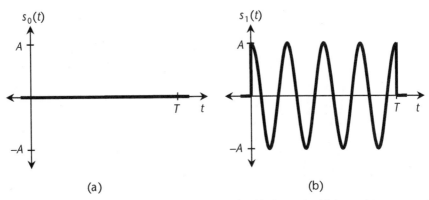

Figure 2.4   Signals for amplitude shift keying. Signal $s_0(t)$ shown in (a) is used to represent a bit value of 0, and $s_1(t)$ shown in (b) is used to represent a 1.

for $m \in \{0,1,\dots, M-1\}$, where f_c is the carrier frequency and ϕ_c is the carrier phase. The amplitude, A_m, depends on the value to be transmitted. The energy required to transmit $s_m(t)$ is

$$E_m = \int_{-\infty}^{\infty} s_m^2(t)\, dt = \frac{A_m^2 T}{2}\left[1 + \frac{\sin 4\pi f_c T}{8\pi f_c T}\right] \cong \frac{A_m^2 T}{2} \tag{2.4}$$

Because the carrier frequency is usually large compared to the symbol duration, $f_c T$ is typically quite large and the approximation is very accurate. The average transmitted energy per signal, assuming that all signals are equally likely to be transmitted, is

$$E_s = \frac{1}{M}\sum_{m=0}^{M-1} E_m \tag{2.5}$$

and the average transmitted energy per bit is $E_b = E_s/\log_2 M$.

Transmission of a long packet of bits is carried out sequentially over multiple time slots. For example, to transmit the sequence of bits 10 11 01 00 11 using ASK with $M = 4$, transmission would be carried out over five time slots, with two bits transmitted per time slot. Using the set of amplitudes $A_m \in \{0, A, 2A, 3A\}$, the transmitted ASK signal would be as shown in Figure 2.5.

2.2.2   Phase Shift Keying (PSK)

Instead of modulating the amplitude of the carrier wave, it is also possible to modulate the phase while keeping the amplitude fixed. This is known as *phase shift keying* (PSK). The transmitted signals are given by

$$s_m(t) = \begin{cases} A_c \cos(2\pi f_c t + \phi_m) & 0 \le t \le T \\ 0 & \text{otherwise} \end{cases} \tag{2.6}$$

for $m \in \{0,1,\dots, M-1\}$, where f_c is the carrier frequency and A_c is the carrier amplitude. The phase, ϕ_m, depends on the value to be transmitted. It is common to use the relationship

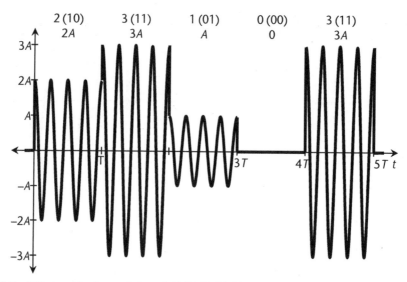

Figure 2.5 ASK signal for transmission of 10 11 01 00 11.

$$\phi_m = \frac{2\pi}{M}m \tag{2.7}$$

for the carrier phase. The average transmitted energy per signal is $E_s = A_c^2 T/2$, which is the same for all m.

When $M = 2$, the scheme is known as *binary phase shift keying* (BPSK), and the two possible transmitted signals are shown in Figure 2.6. Although the two signals look very similar (one is just the negative of the other), it is possible for a well-designed receiver to distinguish between the two signals, even more so than for binary ASK. For the case when $M = 4$, when the scheme is referred to as *quaternary phase shift keying* (QPSK), the transmitted signal used to represent the sequence of bits 10 11 01 00 11 would be as shown in Figure 2.7. We can see that the phase changes with each symbol interval, but the signal amplitude remains constant.

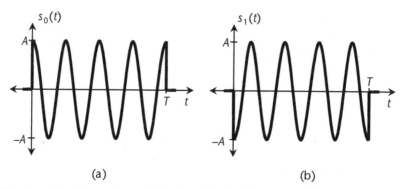

(a)   (b)

Figure 2.6 Signals for binary phase shift keying. Signal $s_0(t)$ shown in (a) is used to represent a bit value of 0, and $s_1(t)$ shown in (b) is used to represent a 1.

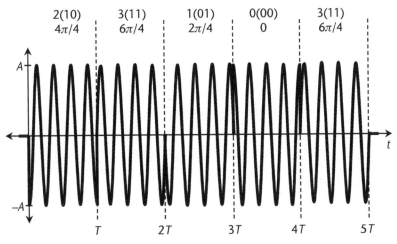

Figure 2.7 QPSK signal for transmitted bit sequence 10 11 01 00 11.

2.2.3 Frequency Shift Keying (FSK)

A third modulation scheme, *frequency shift keying* (FSK), involves keeping the amplitude and phase of the carrier wave constant, while varying the frequency depending on the transmitted data. For $m \in \{0, 1, \ldots, M-1\}$ the transmitted signals are given by

$$s_m(t) = \begin{cases} A_c \cos(2\pi f_m t + \phi_c) & 0 \le t \le T \\ 0 & \text{otherwise} \end{cases} \tag{2.8}$$

The carrier frequency, f_m, depends on the value to be transmitted, and can be related to m by

$$f_m = f_c + m f_\Delta \tag{2.9}$$

where f_c is the nominal carrier frequency and f_Δ is the frequency offset between signals. The frequency offset is typically chosen to be $f_\Delta = 1/T$. Figure 2.8 illustrates

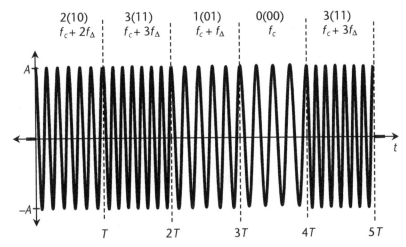

Figure 2.8 FSK signal for transmitted bit sequence 10 11 01 00 11.

the transmitted signal when FSK with $M = 4$ and $f_\Delta = 1/T$ is used to transmit the sequence of bits 10 11 01 00 11. Like PSK, FSK exhibits a constant signal envelope, but because FSK is not spectrally efficient it is seldom used, except for inexpensive low-data-rate applications with $M = 2$.

2.2.4 Quadrature Amplitude Modulation (QAM)

For applications where spectral efficiency is important, *quadrature amplitude modulation* (QAM) is widely used. This modulation scheme involves transmitting two ASK signals simultaneously, one with a cosine wave carrier (the *in-phase* carrier) and one with a sine wave carrier (the *quadrature phase* carrier). The transmitted signal has the form

$$s_m(t) = \begin{cases} A_{I,m} \cos(2\pi f_c t) - A_{Q,m} \sin(2\pi f_c t) & 0 \le t \le T \\ 0 & \text{otherwise} \end{cases} \tag{2.10}$$

where $A_{I,m}$ and $A_{Q,m}$ are the amplitudes of the in-phase and quadrature phase carriers, respectively, for symbol m. Typically the amplitudes are both positive and negative, for example, $A_{I,m}, A_{Q,m} \, \varepsilon \{-3A, -A, A, 3A\}$, as this is more energy efficient. It is worthwhile to note that QAM is a special case where both the amplitude and phase of a carrier are modulated.

2.3 Signal Models

To better understand the operation of communication systems, it is necessary to have some understanding of different mathematical representations of the transmitted signals. These signal models are the complex lowpass equivalent signal representation and the signal space representation. These different models simplify the design and analysis of the system.

2.3.1 Complex Lowpass Equivalent Signal Representation

For modulation schemes that involve modulating the phase and/or the amplitude of the carrier, it is convenient to describe the signals in terms of their complex lowpass equivalent signal representation. The transmitted bandpass signal used to represent symbol $m \, \varepsilon \{0,1,\ldots, M-1\}$ can be described as

$$s_m(t) = \sqrt{E_m} \, b_T(t) \sqrt{2} \cos(2\pi f_c t + \phi_m) \tag{2.11}$$

where E_m is the energy of $s_m(t)$, f_c is the carrier frequency, ϕ_m is the carrier phase, and $b_T(t)$ is the pulse shape. The pulse shape could be a simple rectangular pulse,

$$b_T(t) = \begin{cases} 1/\sqrt{T} & 0 \le t \le T \\ 0 & \text{otherwise} \end{cases} \tag{2.12}$$

or it could be something more elaborate to better control the spectral properties of the transmitted signal as described later in this chapter. The pulse shape should be normalized to have unit energy, that is,

$$\int_{-\infty}^{\infty} |h_T(t)|^2 \, dt = 1 \tag{2.13}$$

Using the identity $\cos(A + B) = \cos A \cos B - \sin A \sin B$, we can rewrite (2.11) as

$$s_m(t) = \sqrt{E_m} \cos\phi_m h_T(t)\sqrt{2}\cos(2\pi f_c t) - \sqrt{E_m}\sin\phi_m h_T\sqrt{2}\sin(2\pi f_c t)$$
$$= s_{I,m}(t)\sqrt{2}\cos(2\pi f_c t) - s_{Q,m}(t)\sqrt{2}\sin(2\pi f_c t) \tag{2.14}$$

where $s_{I,m}(t) = \sqrt{E_m}\cos\phi_m h_T(t)$ is the lowpass signal transmitted on the in-phase (I) carrier, $\sqrt{2}\cos(2\pi f_c t)$, and $s_{Q,m}(t) = \sqrt{E_m}\sin\phi_m h_T(t)$ is the lowpass signal transmitted on the quadrature phase (Q) carrier, $-\sqrt{2}\sin(2\pi f_c t)$.

Alternatively, we can reexpress (2.11) as

$$s_m(t) = \Re\left\{\sqrt{E_m}e^{j\phi_m}h_T(t)\sqrt{2}e^{j2\pi f_c t}\right\} = \left\{\Re\{s_{l,m}(t)\sqrt{2}e^{j2\pi f_c t}\right\} \tag{2.15}$$

where $\Re\{\cdot\}$ denotes the real part and $s_{l,m}(t) = \sqrt{E_m}e^{j\phi_m}h_T(t)$ is the complex lowpass equivalent transmitted signal. Note that $s_{I,m}(t) = s_{I,m}(t) + js_{Q,m}(t)$, so the complex lowpass equivalent signal is merely a compact notation for representing the lowpass signals on the I and Q channels.

2.3.2 Signal Space Diagrams

Another useful way to express the signals, particularly with regards to measuring the difference between two signals, involves the use of signal space diagrams. To understand signal space diagrams, it is first useful to review the key concepts of signal spaces, which are very similar to vector spaces as taught in most introductory linear algebra courses, but are based on analog signals instead of vectors. The most important difference is in the definition of the inner product, which for two signals, $v(t)$ and $w(t)$, is defined as

$$\langle v(t), w(t)\rangle \triangleq \int_{-\infty}^{\infty} v(t)w(t)\, dt \tag{2.16}$$

The definition of the inner product is at the heart of signal space analysis. It is useful to compare the definition of the inner product for signals given by (2.16) with the corresponding definition for vectors. The inner (dot) product of two N-dimensional vectors, $\mathbf{v} = [v_1 v_2 \ldots v_N]$ and $\mathbf{w} = [w_1 w_2 \ldots w_N]$, is

$$\langle \mathbf{v}, \mathbf{w}\rangle \triangleq \sum_{n=1}^{N} v_n w_n \tag{2.17}$$

The main difference is that with signals we integrate over time whereas with vectors we sum over the vector elements.

Example 2.1:   The Inner Product of Two Signals

Find the inner product of $s_1(t)$ and $s_2(t)$ shown in Figure 2.9.

Solution:

From (2.16), the inner product of the two signals is

$$\langle s_1(t), s_2(t) \rangle = \int_{-\infty}^{\infty} s_1(t)s_2(t)\,dt = \int_0^{\frac{T}{2}}(A)(2A)\,dt + \int_{\frac{T}{2}}^{T}(A)(0)\,dt = A^2T \tag{2.18}$$

With the definition of the inner product of two signals given by (2.16), it is now easy to introduce the remaining important signal space concepts. The norm of a signal is defined as

$$\|s(t)\| \triangleq \sqrt{\langle s(t), s(t) \rangle} \triangleq \sqrt{\int_{-\infty}^{\infty} |s(t)|^2\,dt} \tag{2.19}$$

Whereas the norm of a vector is its length, the norm of a signal is the square root of its energy. A signal is said to be *normal* (or *normalized*) if $\|s(t)\| = 1$ (i.e., it has unit energy). Two signals are *orthogonal* if their inner product is zero. All pairs of signals that do not overlap in time will have an inner product of zero and are therefore orthogonal. However, there are many pairs of signals that do overlap and are still orthogonal. Two such signals are shown in the following example.

Example 2.2:   Orthogonal Signals

Show that $s_1(t)$ and $s_2(t)$ shown in Figure 2.10 are orthogonal.

Solution:

To show that two signals are orthogonal, we must show that their inner product is zero:

$$\langle s_1(t), s_2(t) \rangle = \int_0^{\frac{T}{2}}(A)(A)\,dt + \int_{\frac{T}{2}}^{T}(A)(-A)\,dt = A^2\frac{T}{2} - A^2\frac{T}{2} = 0 \tag{2.20}$$

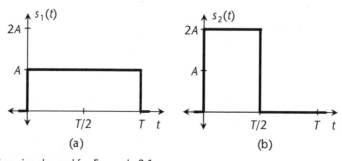

(a)   (b)

Figure 2.9   Two signals used for Example 2.1.

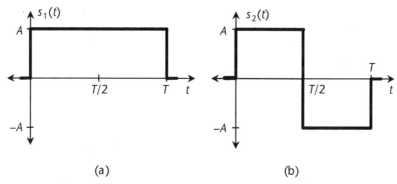

Figure 2.10   Two orthogonal signals used for Example 2.2.

A set of signals are said to be *orthonormal* if they are all normal, and every signal is orthogonal to the other signals in the set. That is, the N signals $\{\phi_n(t)|n \, \varepsilon \, \{1,2, \ldots N\}\}$ are orthonormal if

$$\langle \phi_k(t), \phi_l(t) \rangle = \begin{cases} 1 & \text{if } k = l \\ 0 & \text{if } k \neq l \end{cases} \tag{2.21}$$

for all $k, l \in \{1,2, \ldots N\}$. Any set of orthonormal signals can be used as the basis signals for some signal space. The span of the signal space is the set of all signals that can be expressed as linear combinations of the basis signals.

The set of all of the transmitted signals used in any modulation scheme can always be expressed as linear combinations of a small set of basis signals. In general, any set of M signals can be expressed as linear combinations of $N \leq M$ basis signals. That is, the signals $\{s_m(t)|m \, \varepsilon \, \{1,2, \ldots, M\}\}$ can be expressed in terms of the basis signals $\{\phi_n(t)|n \, \varepsilon \, \{1,2, \ldots, N\}\}$ as

$$s_m(t) = \sum_{n=1}^{N} s_{m,n} \phi_n(t) \tag{2.22}$$

where $s_{m,n}$ is the weight of the contribution of $\phi_n(t)$ to $s_m(t)$. The N-dimensional point $s_m = (s_{m,1}, s_{m,2}, \ldots s_{m,N})$ gives the coordinates of signal $s_m(t)$ in the signal space defined by $\{\phi_n(t)\}$. These vectors can be plotted in a *signal space diagram*, which is an N-dimensional graph using the basis signals as the axes. The set of points, $\{s_m\}$, in the signal space diagram corresponding to the set of signals, $\{s_m(t)\}$, is also known as the *signal constellation*.

Example 2.3:   Signal Space Representation

The four signals shown in Figure 2.11 can be expressed as linear combinations of the two basis signals $\phi_1(t)$ and $\phi_2(t)$ shown in Figure 2.12. To demonstrate this, we observe that $\phi_1(t)$ and $\phi_2(t)$ are valid basis signals because they are orthonormal. They are orthogonal because they do not overlap, so their inner product must be zero, and they are each normalized (the specified amplitude of $\sqrt{2/T}$ ensures this).

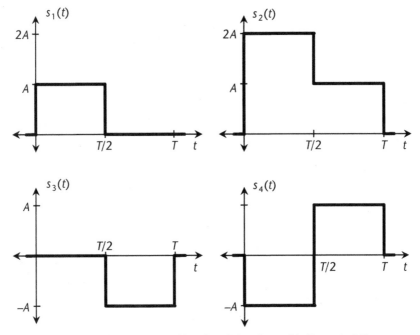

Figure 2.11   The four possible transmitted baseband signals used in Example 2.3.

Furthermore, the data signals can be expressed as linear combinations of the basis signals as follows:

$$s_1(t) = A\sqrt{\frac{T}{2}}\phi_1(t)$$

$$s_2(t) = 2A\sqrt{\frac{T}{2}}\phi_1(t) + A\sqrt{\frac{T}{2}}\phi_2(t)$$

$$(2.23)$$

$$s_3(t) = -A\sqrt{\frac{T}{2}}\phi_2(t)$$

$$s_4(t) = -A\sqrt{\frac{T}{2}}\phi_1(t) + A\sqrt{\frac{T}{2}}\phi_2(t)$$

The four points, $s_1 = \left(A\sqrt{T/2},0\right)$, $s_2 = \left(2A\sqrt{T/2}, A\sqrt{T/2}\right)$, $s_3 = \left(0,-A\sqrt{T/2}\right)$, and $s_4 = \left(-A\sqrt{T/2}, A\sqrt{T/A}\right)$ can be plotted in a signal space diagram as shown in Figure 2.13.

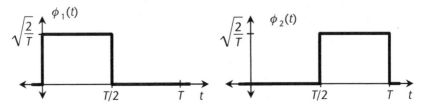

Figure 2.12   The basis signals for Example 2.3.

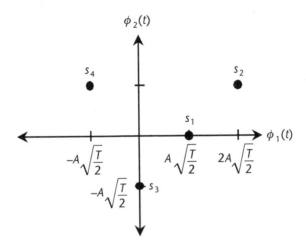

Figure 2.13 Signal space diagram showing the signals of Figure 2.11.

The signal space diagram is a very useful tool for describing the signals. The square of the distance between a signal point and the origin gives the energy of the signal, so, for example, the energy of $s_4(t)$ is

$$\|s_4\|^2 = \left(-A\sqrt{T/A}\right)^2 + \left(A\sqrt{T/A}\right)^2 = A^2 T \tag{2.24}$$

The distance between any two signals is useful for quantifying the similarity between two signals, so two signals that are very close to each other in the signal space diagram are very similar. As we will see later in the chapter, the receiver's ability to distinguish between two signals depends on how close together they are. Ideally we want to select our transmitted signals so they are as far apart as possible. In general, for any modulation scheme we can increase the separation between points merely by increasing the energy of all the transmitted signals. As a result, the probability that the receiver will make an error decreases as the average transmitted energy increases. However, due to physical constraints, there is a limit on the maximum energy that can be transmitted, so for a fixed average energy it is desirable to use a modulation scheme with as large a separation between the points as possible.

Example 2.4: Signal Space Representation

For the four signals shown in Figure 2.11 used in Example 2.3, find the minimum separation between the points. Express the answer in terms of the average signal energy. Suggest a different set of signals in the same signal space that has a larger minimum separation for the same average energy.

Solution:
The closest pair of points are s_1 and s_2 (or s_1 and s_3), and they have a separation of

$$\Delta = \|s_1 - s_2\| = \sqrt{\left(A\sqrt{T/2} - 2A\sqrt{T/2}\right)^2 + \left(0 - A\sqrt{T/2}\right)^2} = \sqrt{A^2 T} \tag{2.25}$$

The average energy is

$$E_s = \frac{1}{M}\sum_{m=1}^{M}\left\|s_m\right\|^2 = \frac{1}{4}\left(A^2\frac{T}{2}+5A^2\frac{T}{2}+A^2\frac{T}{2}+2A^2\frac{T}{2}\right) = \frac{9}{4}A^2\frac{T}{2} \qquad (2.26)$$

so the minimum separation is

$$\Delta = \sqrt{\frac{8}{9}E_s} \qquad (2.27)$$

A better constellation is shown in Figure 2.14. This constellation has a separation of $\Delta = \sqrt{2E_s}$, so for the same average energy, we would expect that a communication system based on the signals given in Figure 2.14 to have a lower probability of error than a system based on Figure 2.13.

We can express the transmitted signals given by (2.11) in a signal space by using

$$\phi_I(t) = h_T(t)\sqrt{2}\cos(2\pi f_c t) \qquad (2.28)$$

and

$$\phi_Q(t) = -h_T(t)\sqrt{2}\sin(2\pi f_c t) \qquad (2.29)$$

as the basis signals. Provided that f_c is sufficiently large compared to the bandwidth of $h_T(t)$, which is typically the case, we note that $\phi_I(t)$ and $\phi_Q(t)$ are normalized (i.e., they each have unit energy) and are orthogonal (i.e., their inner product, $\int_{-\infty}^{\infty}\phi_I(t)\phi_Q(t)\,dt$ is zero), so they are orthonormal and therefore can be used as basis signals. From (2.14) we can express $s_m(t)$ as

$$s_m(t) = \left(\sqrt{E_m}\cos\phi_m\right)\phi_I(t) + \left(\sqrt{E_m}\sin\phi_m\right)\phi_Q(t) = s_{I,m}\phi_I(t) + s_{Q,m}\phi_Q(t) \qquad (2.30)$$

where $s_{I,m} = \sqrt{E_m}\cos\phi_m$ and $s_{Q,m} = \sqrt{E_m}\sin\phi_m$ define the coordinates of $s_m(t)$ in the signal space.

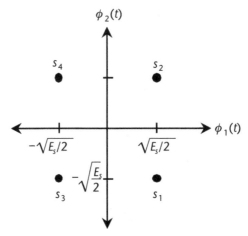

Figure 2.14   An example of a signal constellation with a large separation between the points.

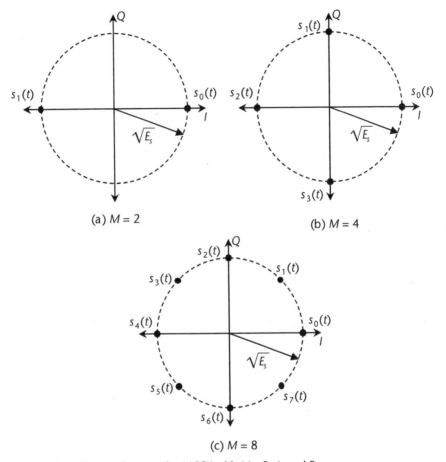

(a) $M = 2$ (b) $M = 4$

(c) $M = 8$

Figure 2.15   Signal space diagrams for M-PSK with $M = 2$, 4, and 8.

For M-ary PSK, where $\phi_m = 2\pi m/M$ and $E_m = E_s = A_c^2 T/2$ is constant for all m, Figure 2.15 shows the signal space diagrams for $M = 2$, 4, and 8. Observe that for M-PSK the points in the signal constellation are evenly space on a circle with radius $\sqrt{E_s}$. Figure 2.16 shows the signal space diagrams for M-ASK where $E_m = m^2 B^2 = m^2 A^2 T/2$ and $\phi_m = 0$ for $M = 2$ and $M = 4$. One attribute of M-ASK signal constellations is that all the points fall in a line. The signal space diagrams for 4-QAM and 16-QAM are shown in Figure 2.17. QAM constellations are typically characterized by having the signal points arranged in a rectangular grid.

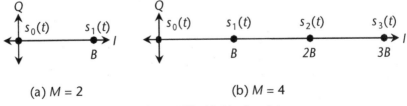

(a) $M = 2$ (b) $M = 4$

Figure 2.16   Signal space diagrams for M-ASK with $M = 2$ and 4.

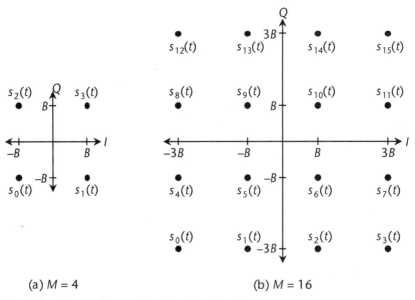

(a) $M = 4$ (b) $M = 16$

Figure 2.17 Signal space diagrams for 4-QAM and 16-QAM.

We can also treat the signal space diagram as a complex plane, and express the signal points as complex numbers, so

$$s_m = s_{I,m} + js_{Q,m} = \sqrt{E_m}\,e^{j\phi_m} \tag{2.31}$$

With this notation, we can see that the square of the distance from the origin to the position of the point in the signal space diagram gives the energy of $s_m(t)$, and the angle to the point, measured counterclockwise from the positive x-axis, gives the phase of the carrier, ϕ_m.

2.4 System Model

We are now able to present a somewhat more detailed model of the communication system. A block diagram is shown in Figure 2.18, which is suitable for any amplitude and/or phase modulation scheme. At the transmitter, the symbol map converts the digital data to be transmitted into the coordinates of the corresponding point in the signal constellation. The pulse shaping filter generates a pulse train based on the desired pulse shape. The modulator converts the complex lowpass equivalent signal into a bandpass signal. The bandpass signal propagates over the wireless channel to the receiver, where a demodulator downconverts the signal back to the baseband. A receive filter blocks out unwanted noise and the signal is sampled at the symbol rate. Based on the samples, a decision device tries to determine the transmitted digital data. Each of the components of the system is described in detail in this section.

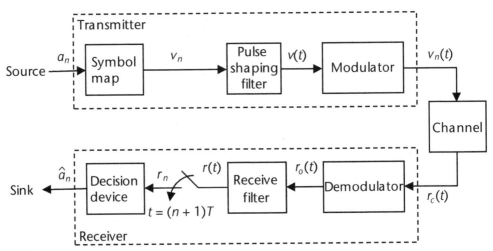

Figure 2.18   Block diagram of a communication system.

2.4.1   Symbol Map

Although we can transmit $\log_2 M$ bits in every symbol, in practice we usually need to transmit much more information. This is achieved by transmitting many symbols sequentially over many subsequent time slots. A data packet (also referred to as a block or a frame) containing N_b bits is partitioned into N_s symbols, each containing $\log_2 M$ bits, so $N_b = N_s \log_2 M$. The packet is expressed as $\mathbf{a} = \left[a_0 a_1 \ldots a_{N_s-1}\right]$, where symbol $a_n \varepsilon \{0,1,\ldots M-1\}$ for $n \varepsilon \{0,1,\ldots N_s-1\}$ represents $\log_2 M$ bits. Depending on the value of the bits, each symbol is mapped to the corresponding point in the signal constellation. If the points in the signal constellation are given by $\{s_m\}$ then the nth transmitted symbol is $v_n = s_{a_n}$.

As an illustrative example, consider the transmission of the data packet $\mathbf{a} = [11\ 01\ 11\ 00\ 10]$ using the 4-ASK signal constellation given in Figure 2.16(b). In this case $M = 4$, $N_b = 10$, $N_s = 5$, and the sequence of transmitted symbols is $\mathbf{v} = [3A, A, 3A, 0, 2A]$.

2.4.2   Pulse-Shaping Filter

The pulse-shaping filter generates a pulse train using each transmitted symbol as the complex-valued amplitude of the desired pulse shape, $h_T(t)$. A new pulse is generated every T seconds, and the transmitted complex lowpass equivalent signal is the pulse train given by

$$v(t) = \sum_{n=0}^{N_s-1} v_n h_T(t - nT) \tag{2.32}$$

For the data packet $\mathbf{a} = [11\ 01\ 11\ 00\ 10]$ transmitted with 4-ASK using the normalized rectangular pulse shape given in (2.12), the transmitted complex lowpass equivalent signal is as shown in Figure 2.19.

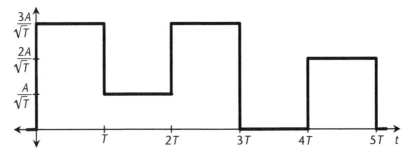

Figure 2.19 Complex lowpass equivalent signal for 4-ASK transmission of **a** = [11 01 11 00 10] with a rectangular pulse shape.

2.4.3 Modulator

The lowpass signal is upconverted to the desired frequency band by modulating the amplitude of a complex carrier, giving the transmitted baseband signal

$$v_c(t) = \Re\{v(t)\sqrt{2}e^{j2\pi f_c t}\} \tag{2.33}$$

Figure 2.20 shows $v_c(t)$ when 4-ASK is used to transmit the message **a** = [11 01 11 00 10].

2.4.4 Additive White Gaussian Noise (AWGN) Channel Model

As the signal propagates from the transmitter to the receiver, it is corrupted in many ways. The most simple channel model is the additive noise channel model, where the only distortion to the signal is the addition of an independent noise signal. The received signal is modeled as

$$r_c(t) = v_c(t) + w_c(t) \tag{2.34}$$

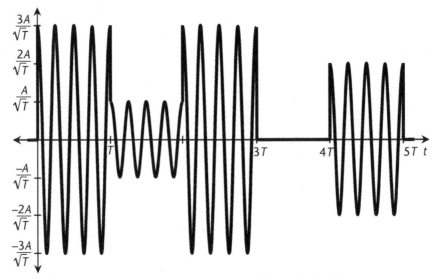

Figure 2.20 Transmitted bandpass 4-ASK signal for **a** = [11 01 11 00 10].

where $v_c(t)$ is the transmitted bandpass data-bearing signal and $w_c(t)$ is the additive noise. Although this channel model is overly simplistic, it is also the most important one to understand, because all the more realistic channel models build up on this basic model.

The additive noise signal is modeled as a white Gaussian random process, and so this channel model is called the additive white Gaussian noise (AWGN) channel model. A sample of the random process at any time, t_1, is a Gaussian random variable, with a mean of zero (i.e., $E[w_c(t_1)] = 0$ for any t_1, where $E[\cdot]$ denotes the expected value). The random process is stationary (that is, its statistics do not change over time), and its power spectral density is constant at $N_0/2$ over all frequencies. The flat (white) PSD implies that two samples of the random process, taken at two different times no matter how close together, are independent.

2.4.5 Demodulator

The received signal is demodulated to the baseband by mixing it with a locally generated carrier reference signal. This signal should have the same frequency and nominal phase as the carrier signal, but with the quadrature phase component reversed in sign. The downconverted signal is

$$r_o(t) = r_c(t)\sqrt{2}e^{-j2\pi f_c t} = v_c(t)\sqrt{2}e^{-j2\pi f_c t} + w_c(t)\sqrt{2}e^{-j2\pi f_c t} \qquad (2.35)$$

By substituting (2.33) for $v_c(t)$ and defining $w_o(t) = w_c(t)\sqrt{2}e^{-j2\pi f_c t}$ as the downconverted noise, we can express $r_o(t)$ as

$$r_o(t) = \{v(t)\sqrt{2}e^{j2\pi f_c t}\}\sqrt{2}e^{-j2\pi f_c t} + w_o(t) \qquad (2.36)$$

By using the identity $\Re\{z\} = \frac{1}{2}z + \frac{1}{2}z^*$ for the real part of a complex number, z, where the superscript $*$ denotes the complex conjugate, we can write (2.36) as

$$r_o(t) = \left[\frac{\sqrt{2}}{2}v(t)e^{j2\pi f_c t} + \frac{\sqrt{2}}{2}v^*(t)e^{-j2\pi f_c t}\right]\sqrt{2}e^{-j2\pi f_c t} + w_o(t)$$
$$= v(t) + v^*(t)e^{-j4\pi f_c t} + w_o(t) \qquad (2.37)$$

The first term in (2.37) is the desired transmitted lowpass signal, and the second term is a high-frequency component that will be removed by the receive filter, which is essentially a lowpass filter. The third term is the downconverted noise, which is still a zero-mean stationary white Gaussian random process, but is complex-valued with a constant PSD of N_0.

2.4.6 Receive Filter

Once the received signal has been downconverted, it must be filtered to remove the high-frequency component and any other signals in adjacent bands. The filter should also emphasize the data-bearing signal while suppressing the noise as much as possible. The optimal filter is one that is matched to the transmitted pulse shape and should therefore have an impulse response of $h_R(t) = h_T(T - t)$. The filtered received signal is

$$r(t) = \int_{-\infty}^{\infty} r_o(t - \tau) h_R(\tau) \, d\tau \tag{2.38}$$

Substituting (2.37) for $r_o(t)$ and ignoring its high-frequency component yields

$$r(t) = \int_{-\infty}^{\infty} v(t - \tau) h_R(\tau) \, d\tau + \int_{-\infty}^{\infty} w_o(t - \tau) h_R(\tau) \, d\tau \tag{2.39}$$

By defining $w(t) = \int_{-\infty}^{\infty} w_o(t - \tau) h_R(\tau) \, d\tau$ as the filtered noise, and substituting (2.32) for $v(t)$, we can express $r(t)$ as

$$
\begin{aligned}
r(t) &= \int_{-\infty}^{\infty} \left[\sum_{n=0}^{N_s-1} v_n h_T(t - \tau - nT) \right] h_R(\tau) \, d\tau + w(t) \\
&= \sum_{n=0}^{N_s-1} v_n \int_{-\infty}^{\infty} h_T(t - \tau - nT) \\
&\quad h_R(\tau) \, d\tau + w(t) = \sum_{n=0}^{N_s-1} v_n h_{TR}(t - nT) + w(t)
\end{aligned} \tag{2.40}
$$

where $h_{TR}(t) = \int_{-\infty}^{\infty} h_T(t - \tau) h_R(\tau) \, d\tau$ is the combined impulse response of the transmit and receive filters. The filtered noise component is still a zero-mean complex Gaussian random process, but it is no longer white.

2.4.7 Signal Sampling

The received signal is sampled once every T seconds starting at time $t = T$. The received samples are given by

$$
\begin{aligned}
r_n = r([n + 1]T) &= \sum_{m=0}^{N_s-1} v_m h_{TR}([n + 1]T - mT) + w([n + 1]T) \\
&= \sum_{m=0}^{N_s-1} v_m h_{TR}([n - m + 1]T) + w_n
\end{aligned} \tag{2.41}
$$

where $w_n = w([n + 1]T)$ is the sampled noise. We see that r_n depends not only on v_n but potentially on all the other transmitted symbols, a phenomenon known as intersymbol interference (ISI). To avoid ISI, it is necessary to choose the pulse carefully. In particular, $h_{TR}(t)$ must have the property that

$$h_{TR}([n - m + 1]T) = \begin{cases} 1 & \text{if } m = n \\ 0 & \text{if } m \neq n \end{cases} \tag{2.42}$$

Any normalized pulse shape, $h_T(t)$, that is nonzero only over the interval $t \, \varepsilon \, [0, T]$ will provide this property when used with a matched filter at the receiver. Some other pulse shapes, such as the impulse response of a root raised-cosine filter, which last longer that T seconds, can also be used and can provide better spectral properties. If the property of (2.42) is met, then (2.41) reduces to

$$r_n = v_n + w_n \tag{2.43}$$

which depends only on the nth transmitted symbol, v_n, corrupted by an additive noise sample, w_n. The noise sample is a zero-mean complex Gaussian random variable with variance $\sigma_w^2 = \frac{1}{2}E[|w_n|^2] = N_0/2$, and the different noise samples are all independent.

The received samples at the output of the matched filter are actually the projection of the received signal onto the signal space of the transmitted signals. To appreciate this interpretation it is useful to view a scatter plot of the received samples. An example, for 4-QAM, is shown in Figure 2.21(a), which was generated by simulating the transmission of a sequence of $N_s = 200$ symbols in the presence of weak additive noise. We can see that the received samples are tightly clustered around the symbol points in the 4-QAM signal constellation. When the noise is stronger, as shown in Figure 2.21(b), the received samples are more dispersed.

2.4.8 Decision Device

The decision device must determine a_n based on the received sample, r_n. Because of the noise introduced in the channel, it is impossible to guarantee that the decision device will always correctly determine what was transmitted. An optimal decision device is one that minimizes the probability of making an error. Theoretically, this can be accomplished by calculating the *a posteriori probabilities* (APPs) that each possible value was sent given the received sample, and then selecting that value which has the highest probability of being sent. This is known as the *maximum a posteriori probability* (MAP) decision rule.

Example 2.5:   MAP Decision Rule

Consider a system where one of the $M = 4$ possible values {0, 1, 2, 3} was transmitted. Suppose the receiver calculates the following four APPs:

$$
\begin{aligned}
\Pr\{0 \text{ sent} \mid r_n \text{ received}\} &= 0.2 \\
\Pr\{1 \text{ sent} \mid r_n \text{ received}\} &= 0.1 \\
\Pr\{2 \text{ sent} \mid r_n \text{ received}\} &= 0.4 \\
\Pr\{3 \text{ sent} \mid r_n \text{ received}\} &= 0.3
\end{aligned}
\tag{2.44}
$$

According to the MAP decision rule, the receiver would decide that 2 was sent, because that is the most likely value to have been transmitted given the received

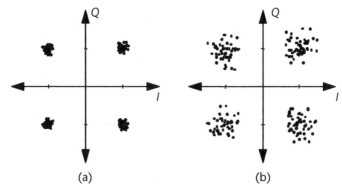

(a)   (b)

Figure 2.21 Scatter plot of received samples for 4-QAM with (a) weak noise and (b) strong noise.

sample. Note that the probability of error in this example is 60%, which is very high, but there is no other decision that the receiver could make that would give a lower probability of error.

In general, the MAP decision rule states that the receiver should choose $\hat{m} = m$ as the transmitted value if

$$\Pr\{m \text{ sent} \mid r_n \text{ received}\} > \Pr\{l \text{ sent} \mid r_n \text{ received}\} \quad \forall l \neq m \qquad (2.45)$$

or equivalently

$$\hat{m} = \arg\max_m \Pr\{m \text{ sent} \mid r_n \text{ received}\} \qquad (2.46)$$

Under the condition that each of the M signals are equally likely to be transmitted, the MAP decision rule can be simplified to the *maximum likelihood* (ML) decision rule, which states the receiver should choose $\hat{m} = m$ if

$$f(r_n \mid m \text{ sent}) > f(r_n \mid l \text{ sent}) \quad \forall l \neq m \qquad (2.47)$$

where $f(r_n \mid m \text{ sent})$ is the likelihood function (the conditional probability density function) of r_n being received given that signal s_m was transmitted. For the AWGN channel, if s_m was transmitted then r_n has a Gaussian distribution with a mean of s_m and a variance of $N_0/2$. The ML decision rule can therefore be written as choosing $\hat{m} = m$ if

$$\frac{1}{\sqrt{\pi N_0}} \exp\left\{ -\frac{|r_n - s_m|^2}{N_0} \right\} > \frac{1}{\sqrt{\pi N_0}} \exp\left\{ -\frac{|r_n - s_l|^2}{N_0} \right\} \quad \forall l \neq m \qquad (2.48)$$

By taking the log of both sides and cancelling terms common to both sides of the inequality, (2.48) reduces to choosing $\hat{m} = m$ if

$$|r_n - s_m|^2 < |r_n - s_l|^2 \quad \forall l \neq m \qquad (2.49)$$

or equivalently

$$\hat{m} = \arg\max_m |r_n - s_m|^2 \qquad (2.50)$$

Since $|r_n - s_m|$ is just the Euclidean distance between the received sample, r_n, and signal s_m in the signal space diagram, the optimal receiver reduces to merely finding the closest possible transmitted signal to r_n in the signal space diagram.

By dividing the signal space into regions corresponding to which signal point is closest, it is possible to further simplify the decision device. For example, for QPSK, as shown in Figure 2.22, the signal space is divided into four regions, denoted by Z_m for $m \in \{0, 1, 2, 3\}$. It is apparent that any point in region Z_m is closer to s_m than to any of the other three signal points, and so if the received sample falls in region Z_m then the receiver should decide that s_m was most likely to have been transmitted. So, for example, if the received sample falls within the left shaded region (region Z_2) in Figure 2.22, the receiver output would be 2. Note that it is theoretically possible, although unlikely, for the received sample to fall directly on a boundary between two decision regions. In this case both symbols are equally likely to have been transmitted,

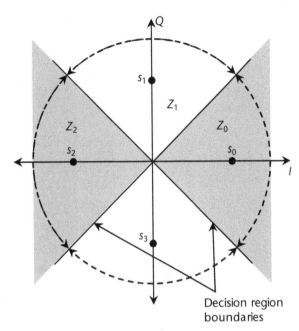

Figure 2.22 QPSK signal constellation with optimal decision regions.

so it does not matter which one the receiver selects. One of the two could be selected randomly, or the decision device could always favor one over the other.

2.5 Probability of Error Analysis

An important measure of the reliability of a communication system is the probability of error. When we use a communication system, we expect the recipient to receive the identical message to what was transmitted, without any errors. Although it is impossible to guarantee error-free communication, we can design systems that have a very low probability that an error will occur. This is accomplished by using many techniques, such as modulation schemes that are more robust against channel noise, error correcting codes with additional transmitted redundancy to correct as many errors as possible, and error detecting codes that can confirm whether or not a message was received correctly and decide whether or not a retransmission of the message is required. In the following we discuss the effects of channel noise on the transmitted signal and illustrate how the noise leads to decision errors. We also derive the probability of error for simple binary communication systems before considering M-ary signaling schemes.

2.5.1 Binary Signaling

Consider the case of binary signalling with signals $s_0(t)$ and $s_1(t)$ used to represent 0 and 1, respectively. These two signals can be represented in a complex signal space diagram by the points $s_0 = s_{I,0} + js_{Q,0}$ and $s_1 = s_{I,1} + js_{Q,1}$ where $s_{I,m}$ and $s_{Q,m}$ and the in-phase and quadrature phase amplitudes of $s_m(t)$, respectively. If the transmitted bit is as likely to be 0 as 1, and transmission takes places over an AWGN channel, the optimal ML decision rule is to choose $\hat{m} = 1$ as the received bit if $|r - s_1|^2 < |r - s_0|^2$, where r is the detector output as a point in the signal space, and the receiver should choose $\hat{m} = 0$ otherwise. That is, if r is closer to s_1 than s_0, then the receiver will assume that the transmitted bit was 1, and will assume that a 0 was transmitted otherwise. Of course, it is always possible that s_0 was transmitted, and the noise was sufficiently large to make r close to s_1, leading to a decision error. We are interested in the probability that this event will occur, along with the probability that s_1 was transmitted, and the noise was sufficiently large to make r close to s_0. We will handle these two cases separately.

The conditional probability of error under the assumption that a 0 is transmitted is simply

$$P_{\varepsilon|0} = \Pr\{1 \text{ received} \mid 0 \text{ sent}\} \tag{2.51}$$

The only way a 1 will be received is if r is closer to s_1 than s_0, so (2.51) can be expressed as

$$P_{\varepsilon|0} = \Pr\{|r - s_1|^2 < |r - s_0|^2 \mid 0 \text{ sent}\} \tag{2.52}$$

If 0 was transmitted, then the received sample, r, will have the form

$$r = s_0 + w \tag{2.53}$$

where w is a complex Gaussian random variable with a mean of zero and a variance of $N_0/2$ (i.e., the real and imaginary parts of w are independent real Gaussian random variables with zero mean and variance $N_0/2$). We can therefore write the conditional probability of error as

$$P_{\varepsilon|0} = \Pr\{|s_0 + w - s_1|^2 < |s_0 + w - s_0|^2\} = \Pr\{|w + s_0 - s_1|^2 < |w|^2\} \tag{2.54}$$

Expressing s_0, s_1, and w in terms of their real and imaginary parts gives

$$\begin{aligned}
P_{\varepsilon|0} = \Pr\Big\{ &(w_I + s_{I,0} - s_{I,1})^2 + (w_Q + s_{Q,0} - s_{Q,1})^2 < w_I^2 + w_Q^2 \Big\} \\
= \Pr\Big\{ &w_I^2 + 2w_I(s_{I,0} - s_{I,1}) + (s_{I,0} - s_{I,1})^2 + w_Q^2 + 2w_Q(s_{Q,0} - s_{Q,1}) \\
&+ (s_{Q,0} - s_{Q,1})^2 < w_I^2 + w_Q^2 \Big\}
\end{aligned} \tag{2.55}$$

Further simplification yields

$$\begin{aligned}
P_{\varepsilon|0} &= \Pr\Big\{ 2w_I(s_{I,0} - s_{I,1}) + 2w_Q(s_{Q,0} - s_{Q,1}) < -(s_{I,0} - s_{I,1})^2 - (s_{Q,0} - s_{Q,1})^2 \Big\} \\
&= \Pr\Big\{ w_I(s_{I,0} - s_{I,1}) + w_Q(s_{Q,0} - s_{Q,1}) < -\frac{1}{2}|s_0 - s_1|^2 \Big\} \\
&= \Pr\Big\{ w_I(s_{I,0} - s_{I,1}) + w_Q(s_{Q,0} - s_{Q,1}) < -\frac{1}{2}d_{01}^2 \Big\}
\end{aligned} \tag{2.56}$$

where d_{01} is simply the distance between s_0 and s_1 in the signal space diagram. For simplicity, define

$$X = w_I(s_{I,0} - s_{I,1}) + w_Q(s_{Q,0} - s_{Q,1}) \tag{2.57}$$

Because X is a linear combination of two Gaussian random variables, w_I and w_Q, X is also Gaussian. It has a mean of zero and a variance of $\frac{N_0}{2} d_{01}^2$, so the conditional probability density function of X is

$$f_X(x) = \frac{1}{\sqrt{2\pi\left(\frac{N_0}{2} d_{01}^2\right)}} \exp\left\{-\frac{x^2}{2\left(\frac{N_0}{2} d_{01}^2\right)}\right\} \tag{2.58}$$

The probability of error can then be solved as

$$P_{\varepsilon|0} = \Pr\{X < -\tfrac{1}{2}d_{01}^2\} = \int_{-\infty}^{-\frac{1}{2}d_{01}^2} f_X(x)\, dx = \int_{-\infty}^{-\frac{1}{2}d_{01}^2} \frac{1}{\sqrt{\pi N_0 d_{01}^2}} \exp\left\{-\frac{x^2}{N_0 d_{01}^2}\right\} dx \tag{2.59}$$

Performing a change of the variable of integration to

$$u = \frac{-x}{\sqrt{N_0 d_{02}^2}} \tag{2.60}$$

yields

$$P_{\varepsilon|0} = \int_{\frac{d_{01}^2}{2\sqrt{N_0 d_{01}^2}}}^{\infty} \frac{1}{\sqrt{\pi}} e^{-u^2}\, du = \frac{1}{2}\,\mathrm{erfc}\left(\frac{d_{01}^2}{2\sqrt{N_0 d_{01}^2}}\right) = \frac{1}{2}\,\mathrm{erfc}\left(\frac{d_{01}}{2\sqrt{N_0}}\right) \tag{2.61}$$

where

$$\mathrm{erfc}(x) = \frac{2}{\sqrt{\pi}}\int_x^{\infty} e^{-u^2}\, du \tag{2.62}$$

is the complementary error function, which measures the area under the tails of the normalized Gaussian probability density function, and can easily be evaluated numerically.

To proceed with the probability of error analysis, we turn our attention to the case where a one is transmitted. In this case an error will occur only if r is closer to s_0 than s_1, so the probability of error can be expressed as

$$P_{\varepsilon|1} = \Pr\left\{\left|r - s_0\right|^2 < \left|r - s_1\right|^2 \,\middle|\, 1 \text{ sent}\right\} \tag{2.63}$$

By performing analysis similar to that given above, it is easy to show that

$$P_{\varepsilon|1} = \frac{1}{2}\,\mathrm{erfc}\left(\frac{d_{01}}{2\sqrt{N_0}}\right) \tag{2.64}$$

To find the probability of error, we average the results from the two cases, giving

$$P_\varepsilon = P_{\varepsilon|0}\,\mathrm{Pr}\{0\ \text{sent}\} + P_{\varepsilon|1}\,\mathrm{Pr}\{1\ \text{sent}\} = \frac{1}{2}P_{\varepsilon|0} + \frac{1}{2}P_{\varepsilon|1} = \frac{1}{2}\mathrm{erfc}\left(\frac{d_{01}}{2\sqrt{N_0}}\right) \qquad (2.65)$$

From this equation we can make some important observations about the probability of error behavior of binary communication systems. Because the complementary error function decreases as its argument increases, we can see that P_ε decreases as the distance, d_{01}, between the two points in the signal space diagram increases, and P_ε increases as the noise power spectral density, N_0, increases. To ensure reliable communication, it is therefore desirable to keep the noise power as low as possible, and to use signals as far apart as possible in the signal space diagram. As far as the probability of error is concerned, neither the actual shapes of the transmitted signals nor their absolute positions in the signal space diagram matter, only their relative positions. Of course, it is always possible to increase the separation between the signals by increasing the energy of the transmitted signals. However, doing so increases the battery drain in portable devices and increases interference with other users. To keep the energy use as low as possible for a given separation between signals, it is better to use signals whose points are centered about the origin of the signal space diagram.

When the binary ASK (BASK) signal constellation shown in Figure 2.16(a) is used for transmission, the separation between the signal points is $d_{01} = B$, so the probability of error from (2.65) is

$$P_{\varepsilon,BASK} = \frac{1}{2}\mathrm{erfc}\left(\frac{B}{2\sqrt{N_0}}\right) \qquad (2.66)$$

The probability of error for digital communication systems is normally expressed in terms of the average transmitted energy per bit (E_b) to simplify comparison between different systems. For binary modulation schemes, only 1 bit is transmitted per symbol, so $E_b = E_s$, which for BASK is $E_s = \frac{1}{2}(0)^2 + \frac{1}{2}(B)^2 = \frac{1}{2}B^2$. This implies that $B = \sqrt{2E_b}$, so the probability of error can be expressed in terms of E_b as

$$P_{\varepsilon,BASK} = \frac{1}{2}\mathrm{erfc}\left(\sqrt{\frac{E_b}{2N_0}}\right) \qquad (2.67)$$

However, for the binary PSK (BPSK) constellation shown in Figure 2.15(a), the separation between the two points is $d_{01} = 2\sqrt{E_s}$ and the energy per bit is $E_b = \frac{1}{2}\left(\sqrt{E_s}\right)^2 + \frac{1}{2}\left(-\sqrt{E_s}\right)^2 = E_s$ so the probability of error in terms of E_b is

$$P_{\varepsilon,BASK} = \frac{1}{2}\mathrm{erfc}\left(\sqrt{\frac{E_b}{N_0}}\right) \qquad (2.68)$$

For binary FSK, when the frequency offset, f_Δ, is chosen to be a multiple of $1/2T$ so the two data signals are orthogonal, the two data signals (appropriately normalized) can serve as the basis signals for the signal space. The coordinates of the symbols are $s_0 = \left(A_c\sqrt{T/2}, 0\right)$ and $s_1 = \left(0, A_c\sqrt{T/2}\right)$ so the separation between the two points is $d_{01} = A_c\sqrt{T}$ and the energy per bit is $E_b = A_c^2 T/2$. Therefore $d_{01} = \sqrt{2E_b}$ so

$$P_{\varepsilon, BFSK} = \frac{1}{2}\,\mathrm{erfc}\left(\sqrt{\frac{E_b}{2N_0}}\right) \tag{2.69}$$

which is the same as for BASK.

The probabilities of error for BASK and BPSK are plotted in Figure 2.23. These plots, known as waterfall curves, show the probability of error on a logarithmic scale as a function of the ratio E_b/N_0 expressed in decibels. We can see that the BER for both schemes diminishes as E_b/N_0 increases (that is, as the transmitted energy, E_b, increases or the noise PSD, N_0, decreases). However, at any given E_b/N_0, the BER for BASK is higher than for BPSK, so BPSK is more robust against errors due to AWGN. Alternatively, to achieve the same target BER, BASK requires higher transmitted energy than BPSK. For example, to achieve a target BER of 10^{-5}, BPSK requires $E_b/N_0 \geq 9.5$ dB whereas BASK requires $E_b/N_0 \geq 12.5$ dB, so BASK requires 3 dB more (i.e., twice as much) transmitted energy than BPSK.

As we will see throughout this section, the probability of error is usually expressed in terms of the ratio of the average energy per bit (E_b) to the noise power spectral density (N_0). From an RF system design perspective, however, the *signal-to-noise ratio* (SNR) is more commonly used. The SNR, γ_s, is the ratio of the signal power, P_S, to the noise power in a given bandwidth, P_N, so $\gamma_s = P_S/P_N$. Because one symbol, with average energy E_s, is transmitted every T seconds, we have $P_S = E_s/T$. The power of the noise, with a power spectral density of N_0, in a bandwidth of W Hz is $P_N = N_0 W$. If we measure the noise power over a bandwidth of $W = 1/T$ Hz, then the SNR is

$$\gamma_s = \frac{P_S}{P_N} = \frac{E_s/T}{N_0 W} = \frac{E_s}{N_0} = \frac{E_b \log_2 M}{N_0} = \gamma_b \log_2 M \tag{2.70}$$

where $\log_2 M$ is the number of bits per symbol and $\gamma_b = E_b/N_0$ is the SNR per bit.

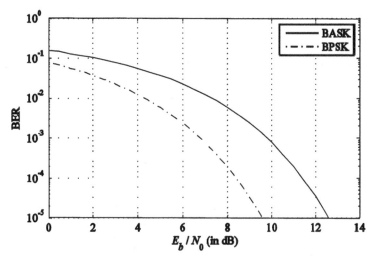

Figure 2.23   BER of BASK and BPSK.

2.5.2 *M*-ary Signaling

For *M*-ary signaling schemes, when there are more than two points in the signal constellation, finding an expression for the probability of error is more difficult. As an illustrative example, we shall consider 4-ASK. The signal constellation, with the optimal decision regions, is shown in Figure 2.24. The binary values associated with each symbol are also shown. To find the probability of error, we consider each transmitted symbol separately. When 00 is sent an error will be made if the received sample is greater than $B/2$, so

$$P_{\varepsilon|00} = \Pr\{r > B/2 \mid 00 \text{ sent}\} = \Pr\{w > B/2\} = \int_{B/2}^{\infty} \frac{1}{\sqrt{\pi N_0}} \exp\left\{-\frac{w^2}{N_0}\right\} dw$$

$$= \frac{1}{2}\operatorname{erfc}\left(\frac{B}{2\sqrt{N_0}}\right)$$

(2.71)

When 01 is transmitted an error will occur if the received sample is either less than $B/2$ or greater than $3B/2$, so

$$P_{\varepsilon|01} = \Pr\{r < B/2 \mid 01 \text{ sent}\} + \Pr\{r > 3B/2 \mid 01 \text{ sent}\}$$

$$= \Pr\{w < -B/2\} + \Pr\{w > B/2\} = \operatorname{erfc}\left(\frac{B}{2\sqrt{N_0}}\right)$$

(2.72)

By similar reasoning

$$P_{\varepsilon|10} = \operatorname{erfc}\left(\frac{B}{2\sqrt{N_0}}\right)$$

(2.73)

and

$$P_{\varepsilon|11} = \frac{1}{2}\operatorname{erfc}\left(\frac{B}{2\sqrt{N_0}}\right)$$

(2.74)

The average probability of error can then be found by taking the average of these four expressions:

$$P_{\varepsilon} = \frac{1}{4}\sum_{m=0}^{3} P_{\varepsilon|m} = \frac{3}{4}\operatorname{erfc}\left(\frac{B}{2\sqrt{N_0}}\right)$$

(2.75)

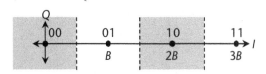

Figure 2.24 Signal space diagram for 4-ASK with decision regions.

Because

$$E_\varepsilon = \frac{1}{4} \sum_{m=0}^{3} (mB)^2 = \frac{1}{4}(0+1+4+9)B^2 = \frac{7}{2}B^2 \tag{2.76}$$

and $E_b = E_s/2$ we have

$$P_\varepsilon = \frac{3}{4}\text{erfc}\left(\sqrt{\frac{E_b}{7N_0}}\right) \tag{2.77}$$

Similar analysis can be used to find the probability of error for higher values of M, and for the other modulation schemes.

This analysis is useful for finding the probability of making a symbol error (that is, the probability that the received symbol differs from the transmitted one). However, we are often more interested in finding the probability of a bit error, also known as the *bit error rate* (BER). For binary systems, where only one bit is transmitted at a time, the BER is the same as the probability of a symbol error. Finding the BER for M-ary modulation schemes is both more tedious and more difficult. Precise calculation of the BER involves first determining the probability that each different symbol is received for every possible transmitted symbol, and then combining these probabilities, weighted according to the number of bit errors that would occur in each case. The conditional probability of a bit error given that symbol m was transmitted is given by

$$P_{b|m} = \sum_{l=0}^{M-1} P_{l|m} \times \left(\frac{\text{\# of bit positions in which } m \text{ and } l \text{ differ}}{\text{\# of bits per symbol}}\right) \tag{2.78}$$

where $P_{l|m} = \Pr\{l \text{ received} \mid m \text{ sent}\}$ is the *transition probability*. The BER is then given by

$$P_b = \frac{1}{M} \sum_{m=0}^{M-1} P_{b|m} \tag{2.79}$$

which is just the average of the conditional probabilities.

Example 2.6: BER for 4-ASK

For example, for the signal space diagram shown in Figure 2.24, the probability that 01 is received given that 11 was transmitted is

$$\begin{aligned}
P_{01|11} &= \Pr\left\{\frac{B}{2} < r < \frac{3B}{2}\middle| 11 \text{ sent}\right\} \\
&= \Pr\left\{\frac{B}{2} < r < \frac{3B}{2}\middle| r = 3B + w\right\} \\
&= \Pr\left\{-\frac{5B}{2} < w < -\frac{3B}{2}\right\} \\
&= \frac{1}{2}\text{erfc}\left(\frac{3B}{2\sqrt{N_0}}\right) - \frac{1}{2}\text{erfc}\left(\frac{5B}{2\sqrt{N_0}}\right)
\end{aligned} \tag{2.80}$$

and if this event were to occur, one of the two transmitted bits would be received incorrectly and the other would be received correctly. Considering the other cases when 11 is transmitted, the conditional probability of bit error is

$$P_{b|11} = \frac{2}{2}(Q_5) + \frac{1}{2}(Q_3 - Q_5) + \frac{1}{2}(Q_1 - Q_3) + \frac{0}{2}(1 - Q_1)$$
$$= \frac{1}{2}Q_1 + \frac{1}{2}Q_5 \tag{2.81}$$

where

$$Q_n = \frac{1}{2}\text{erfc}\left(\frac{nB}{2\sqrt{N_0}}\right) \tag{2.82}$$

has been defined for notational simplicity. By repeating this calculation for the cases where 00, 10, and 11 were transmitted, we find

$$P_{b|10} = P_{b|11} = \frac{1}{2}Q_1 + \frac{1}{2}Q_5 \tag{2.83}$$

and

$$P_{b|01} = P_{b|10} = \frac{3}{2}Q_1 - \frac{1}{2}Q_3 \tag{2.84}$$

Averaging the results gives the BER:

$$P_b = Q_1 - \frac{1}{4}Q_3 + \frac{1}{4}Q_5$$
$$= \frac{1}{2}\text{erfc}\left(\frac{B}{2\sqrt{N_0}}\right) - \frac{1}{8}\text{erfc}\left(\frac{3B}{2\sqrt{N_0}}\right) + \frac{1}{8}\text{erfc}\left(\frac{5B}{2\sqrt{N_0}}\right) \tag{2.85}$$

For large values of B or small values of N_0, the first term is in (2.85) is much bigger than the others, so

$$P_b \cong \frac{1}{2}\text{erfc}\left(\frac{B}{2\sqrt{N_0}}\right) \tag{2.86}$$

Under normal system operation, when the probability of error is small, when errors do occur it is usually because a symbol immediately adjacent to the transmitted symbol is erroneously selected. It is very unlikely that the received sample will fall within the decision region of a more distant symbol. To ensure the BER is small it is therefore a good idea to ensure that adjacent symbols differ in only one bit position, so that when a symbol error occurs, usually only a single bit is received incorrectly. This is achieved by carefully mapping the transmitted bits to the signals using a technique known as *Gray mapping*. For example, with 4-ASK, we should map the bits 11 to the signal at position $2B$ in the signal space diagram and bits 10 to the signals at position $3B$ instead of the other way around, as shown in Figure 2.25. With this Gray mapping we can see that any error into an adjacent symbol region will cause only a single bit error. We note that it is quite easy to find

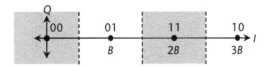

Figure 2.25   Signal space diagram for 4-ASK with Gray mapping.

a suitable Gray mapping for most common modulation schemes, with the exception of M-ary FSK. When Gray mapping is used, the probability of a bit error is well-approximated by

$$P_b \cong \frac{P_\varepsilon}{\log_2 M} \tag{2.87}$$

where P_ε is the probably of a symbol error and $\log_2 M$ is the number of bits per symbol.

2.5.3   BER Comparison of Different Modulation Schemes

In this section we will present formulas for the BER for the most common modulation schemes. Derivations of most of these results can be found in [1]. As shown earlier, the BER for BPSK is

$$P_{b,BPSK} = \frac{1}{2}\mathrm{erfc}\left(\sqrt{\gamma_b}\right) \tag{2.88}$$

where $Y_b = E_b/N_0$ is the SNR per bit. When Gray mapping is used with QPSK, the BER is

$$P_{b,QPSK} = \frac{1}{2}\mathrm{erfc}\left(\sqrt{\gamma_b}\right) \tag{2.89}$$

and in general for larger values of M the BER is

$$P_{b,M-PSK} \cong \frac{1}{\log_2 M}\mathrm{erfc}\left(\sqrt{\gamma_b \log_2 M}\sin\frac{\pi}{M}\right) \tag{2.90}$$

A comparison of the BER for the different M-PSK schemes is shown in Figure 2.26. We observe that the required E_b/N_0 to achieve a target BER increases as M increases, and each doubling of M (i.e., each additional bit transmitted at a time) requires a larger increase in E_b/N_0. This is because for a fixed E_b, the points in the signal constellation must be placed closer together as M increases, so it is more difficult for the receiver to distinguish between the different signals in the presence of noise.

Example 2.7:   SNR to Achieve a Target BER

Suppose we want to determine the minimum SNR required to achieve a target BER of 10^{-3} when BPSK is used. From Figure 2.26 to achieve an error rate of 10^{-3} for BPSK we need E_b/N_0 greater than about 6.8 dB. Because only one bit is transmitted per symbol, the minimum SNR is also 6.8 dB. If 8-PSK is used instead, then a bit error rate of 10^{-3} would require $E_b/N_0 > 10$ dB, and because there are three bits

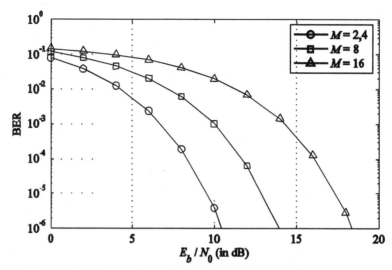

Figure 2.26 Probability of bit error for M-PSK.

per symbol, the required SNR is $10 + 10\log_{10}3 = 14.8$ dB. Thus, 8-PSK requires an increase of 8 dB in the signal power to achieve the same BER as BPSK, which must be balanced with the benefits of transmitting bits three times faster.

The BER of BASK is

$$P_{b,BASK} = \frac{1}{2}\text{erfc}\left(\sqrt{\gamma_b/2}\right)$$

(2.91)

and for 4-ASK it is

$$P_{b,4-ASK} = \frac{1}{2}\text{erfc}\left(\sqrt{\gamma_b/7}\right) - \frac{1}{8}\text{erfc}\left(3\sqrt{\gamma_b/7}\right) + \frac{1}{8}\text{erfc}\left(5\sqrt{\gamma_b/7}\right)$$

(2.92)

while for larger M where M is a power of 2 it is approximated by

$$P_{b,M-ASK} \cong \frac{M-1}{M\log_2 M}\text{erfc}\left(\sqrt{\frac{3\log_2 M}{4(M-1)(M-\frac{1}{2})}\gamma_b}\right)$$

(2.93)

The performance of ASK is shown in Figure 2.27. We observe that ASK exhibits similar behavior to PSK as M increases, but for any given value of M we note that ASK gives worse performance than PSK. This is because the average energy for ASK is much higher than for PSK for a given separation between the points. It is worth mentioning that a more commonly used ASK scheme involves centering the points about the origin in the signal space diagram, thereby reducing the average power, but even with this change ASK is still worse than PSK in terms of the SNR required to achieve a target BER, by about 5 dB.

The BER of 4-QAM is the same as BPSK,

$$P_{b,4-QAM} = \frac{1}{2}\text{erfc}\left(\sqrt{\gamma_b}\right)$$

(2.94)

and for square QAM constellations with larger M (where M is an even power of 2), the BER is approximately

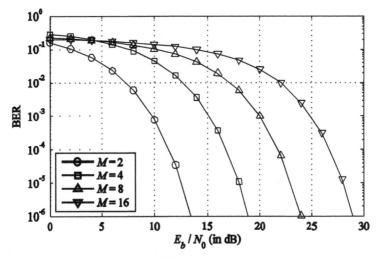

Figure 2.27 Probability of bit error for *M*-ASK.

$$P_{b,M-QAM} \cong \frac{2}{\log_2 M}\left(1 - \frac{1}{\sqrt{M}}\right)\text{erfc}\left(\sqrt{\frac{3}{2(M-1)}\gamma_b \log_2 M}\right) \tag{2.95}$$

The performance of QAM is shown in Figure 2.28. Although the performance degrades as *M* increases, it does so much more slowly than the previous two modulation schemes. In fact 64-QAM gives exactly the same performance as 8-ASK, and in general QAM gives twice the throughput as ASK for the same BER, when compared to an ASK scheme with the signal points centered about the origin, as they are for QAM.

The BER of BFSK is

$$P_{b,BFSK} = \frac{1}{2}\text{erfc}\left(\sqrt{\gamma_b/2}\right) \tag{2.96}$$

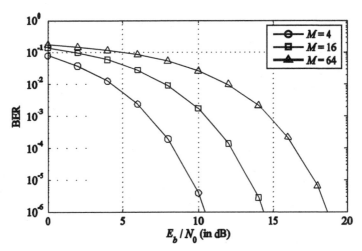

Figure 2.28 Probability of bit error for *M*-QAM.

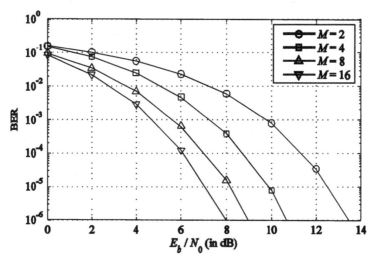

Figure 2.29 Probability of bit error for M-FSK.

and for larger M is approximated by

$$P_{b,M-FSK} \cong \frac{M}{4} \text{erfc}\left(\sqrt{\frac{1}{2}\gamma_b \log_2 M} \right) \tag{2.97}$$

As can be seen in Figure 2.29, the performance of FSK is somewhat different, in that as M increases, the performance actually improves. That is, it is better to transmit more bits at a time as this leads to a lower probability of error, completely the opposite of what occurs with the other modulation schemes. However, this is not really a fair comparison, because the bandwidth required to transmit the signals increases as M increases with FSK, while it remains constant with the other schemes. Put another way, to keep a fixed signal bandwidth while increasing M with FSK, it is necessary to transmit data very slowly (i.e., the symbol duration, T, must be made very long). As FSK is very spectrally inefficient, it is seldom used except for with $M = 2$, when a very simple receiver architecture can be used.

2.6 Signal Spectral Density

The probability of error is an important parameter of a communication system because it measures the reliability of a link. If the probability of error is too high, the need for retransmission increases, thereby reducing the system throughput. An equally important measure is the spectral efficiency of the system, as measured in part by the power spectral density of the transmitted signal. The PSD is also important for ensuring the transmitted signal meets the regulatory requirements to prevent interference with other wireless users.

Provided that the transmitted symbols are independent and all symbols are equally likely to occur, the spectrum of the lowpass signal, $v(t)$, is given by

$$\Phi_v(f) = \frac{\sigma_v^2}{T}|H_T(f)|^2 + \frac{|\mu_v|^2}{T^2}\sum_{m=-\infty}^{\infty}\left|H_T\left(\frac{m}{T}\right)\right|^2 \delta\left(f - \frac{m}{T}\right) \tag{2.98}$$

where

$$H_T(f) = \int_{-\infty}^{\infty} h_T(t) e^{-j2\pi ft}\, dt \tag{2.99}$$

is the Fourier transform of $h_T(t)$, $\delta(t)$ is the Dirac delta (impulse) function,

$$\mu_v = \frac{1}{M} \sum_{m=0}^{M-1} s_m \tag{2.100}$$

is the mean of the points in the signal constellation, and

$$\sigma_v^2 = \frac{1}{M} \sum_{m=0}^{M-1} |s_m|^2 - |\mu_v|^2 = E_S - |\mu_v|^2 \tag{2.101}$$

is the variance. The spectrum of the bandpass signal, $v_c(t)$, can be expressed in terms of the spectrum of the lowpass signal as

$$\Phi_{v_c}(f) = \frac{1}{2} \Phi_v(f - f_c) + \frac{1}{2} \Phi_v^*(f + f_c) \tag{2.102}$$

Example 2.8: Binary ASK with a Rectangular Pulse

The Fourier transform of the normalized rectangular pulse is

$$H_T(f) = \sqrt{T}\, \frac{\sin \pi f T}{\pi f T} e^{-j\pi fT} \tag{2.103}$$

The points in the signal constellation shown in Figure 2.16 are $s_0 = 0$ and $s_1 = B$. Based on the signal constellation, we calculate $\mu_v = B/2$, $E_s = B^2/2$, and $\sigma_v^2 = B^2/4$. Using (2.98), the spectrum of the lowpass signal is therefore

$$\Phi_v(f) = \frac{B^2/4}{T} \left| \sqrt{T}\, \frac{\sin \pi f T}{\pi f T} e^{-j\pi fT} \right|^2 + \frac{|B/2|^2}{T^2} \sum_{m=-\infty}^{\infty} \left| \sqrt{T}\, \frac{\sin \pi \frac{m}{T} T}{\pi \frac{m}{T} T} e^{-j\pi \frac{m}{T} T} \right|^2 \delta\left(f - \frac{m}{T}\right)$$

$$= \frac{E_s}{2} \left(\frac{\sin \pi f T}{\pi f T} \right)^2 + \frac{E_s}{2T} \sum_{m=-\infty}^{\infty} \left(\frac{\sin \pi m}{\pi m} \right)^2 \delta\left(f - \frac{m}{T}\right) \tag{2.104}$$

Because, for integral m,

$$\frac{\sin \pi m}{\pi m} = \begin{cases} 1 & \text{if } m = 0 \\ 0 & \text{if } m \neq 0 \end{cases} \tag{2.105}$$

this reduces to

$$\Phi_v(f) = \frac{E_s}{2} \left(\frac{\sin \pi f T}{\pi f T} \right)^2 + \frac{E_s}{2T} \delta(f) \tag{2.106}$$

which is shown in Figure 2.30. We can see that the bulk of the signal power is contained in the main lobe centered around 0 Hz, but there is also some signal power contained in the sidelobes at higher frequencies. There is also an impulse at 0 Hz, which is due to the dc offset of the signals (i.e., the signal points are not centered about the origin in the ASK signal constellation, so $\mu_v \neq 0$). The PSD of the transmitted

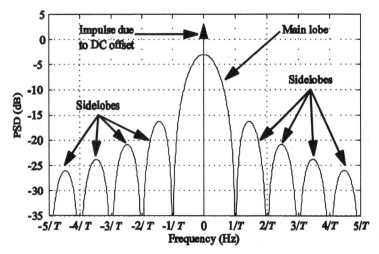

Figure 2.30   PSD of baseband signal used with ASK.

bandpass signal, as given by (2.102), is similar, except it is centered around $\pm f_c$, and is scaled by a factor of 1/2.

Finding an exact expression for the PSD of FSK is more difficult, but a good approximation can be found by noting that FSK can be viewed as the superposition of M different ASK signals, each with a different carrier frequency, where only one ASK signal is transmitted at a time. For example, the PSD of binary FSK with carrier frequencies of $f_0 = 10/T$ and $f_1 = 11/T$, so the frequency separation is $f_\Delta = 1/T$ is as shown in Figure 2.31.

2.6.1   Signal Bandwidth

Because the useful wireless spectrum is a limited resource that is shared by every user, it is important to understand how much of the frequency band is required to transmit a signal. Formally, the bandwidth of a signal is defined as the size of the range of frequencies for which the PSD is nonzero. However, this definition is of

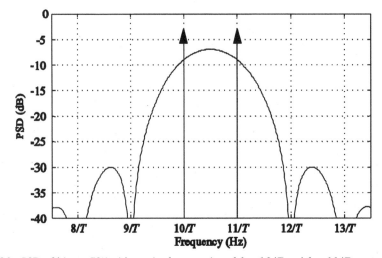

Figure 2.31   PSD of binary FSK with carrier frequencies of $f_0 = 10/T$ and $f_1 = 11/T$.

little practical use, as, strictly speaking, all practical signals have infinite bandwidth. Nonetheless, the PSD of most signals is essentially nonzero over only a finite range of frequencies, and there are several practical definitions of bandwidth that capture this reality. These include the null-to-null bandwidth, the 3-dB bandwidth, and the fractional power bandwidth.

To illustrate these definitions, we consider the PSD of 4-QAM as an example. The PSD of 4-QAM with a rectangular pulse shape is found according to (2.98) and (2.102) as

$$\Phi_{v_c}(f) = \frac{E_s}{2}\left(\frac{\sin \pi[f - f_c]T}{\pi[f - f_c]T}\right)^2 + \frac{E_s}{2}\left(\frac{\sin \pi[f + f_c]T}{\pi[f + f_c]T}\right)^2 \qquad (2.107)$$

This is shown in Figure 2.32, where $f_c = 10/T$ and $E_s = 1$. The null-to-null bandwidth is the width of the main lobe (that is, the separation between the closest spectral nulls to the left and right of the carrier frequency. In this example the null-to-null bandwidth is $2/T$. The 3-dB bandwidth is the separation between the points to the left and right of the carrier frequency where the PSD drops 3 dB below its peak value, so in this case the 3-dB bandwidth is approximately $0.89/T$. The fractional power bandwidth gives the range of frequencies which contain a specific percentage of the total signal power. This is typically computed numerically, and in this example the 99% power bandwidth is roughly $20/T$. In practice, the null-to-null bandwidth is most widely used to quickly characterize the amount of spectrum occupied by the signal.

2.6.2 Pulse Shaping and Intersymbol Interference

To make efficient use of the spectrum, it is desirable to use a bandwidth as small as possible for a given symbol transmission rate, $1/T$, while using a pulse shape that has very low sidelobes. However, if the pulse duration is limited to T seconds, it is not possible to have both low sidelobes and a null-to-null bandwidth of less than $1/T$. As a solution, it is possible to use pulse shapes that are longer than T seconds,

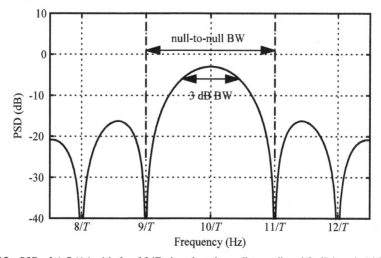

Figure 2.32 PSD of 4-QAM with $f_c = 10/T$, showing the null-to-null and 3-dB bandwidths.

while still transmitting new signals every T seconds. Of course, this will result in signals that overlap in time. Each transmitted symbol may interfere with the next few symbols, resulting in a phenomenon known as *intersymbol interference* (ISI). Although the effects of this interference can be mitigated by using equalization techniques at the receiver, these techniques either substantially increase the complexity of the receiver, or yield inferior BER performance. However, not all longer pulse shapes result in ISI. In fact, the only condition on the pulse shape to avoid ISI is that

$$h_{TR}(nT) = \begin{cases} K & \text{if } n = c \\ 0 & \text{if } n \neq c \end{cases} \tag{2.108}$$

for some real constant K and integer constant c, where $h_{TR}(t)$ is the combined impulse response of the transmit pulse shaping filter and the receiver's matched filter. According to the Nyquist condition for zero ISI, (2.108) is valid as long as $h_{TR}(f)$ has the property that

$$\sum_{n=-\infty}^{\infty} H_{TR}\left(f - \frac{n}{T}\right) \tag{2.109}$$

is constant for all f. It can be shown that this property cannot be met with any pulse shape with a bandwidth of less than $1/T$, so $1/T$ is the minimum bandwidth needed to prevent ISI.

Some popular pulse shapes that meet the Nyquist condition are those based on a raised cosine representation in the frequency domain,

$$H_{TR}(f) = \begin{cases} T & \text{for } |f| < \dfrac{1-\alpha}{2T} \\ T\cos^2\left[\dfrac{\pi T}{2\alpha}\left(|f| - \dfrac{1-\alpha}{2T}\right)\right] & \text{for } \dfrac{1-\alpha}{2T} \leq |f| \leq \dfrac{1+\alpha}{2T} \\ 0 & \text{for } |f| > \dfrac{1+\alpha}{2T} \end{cases} \tag{2.110}$$

where α is known as the *roll-off factor*. Some examples are shown in Figure 2.33 for different values of α. When $\alpha = 0$, the signal has a bandwidth of exactly $1/T$. By taking the inverse Fourier transform of the square root of $H_{TR}(f)$, we can find the transmitted pulse shape in the time domain,

$$h_T(t) = \frac{\sqrt{T}}{1 - \left(\dfrac{4\alpha t}{T}\right)^2}\left[\frac{\sin 2\pi\left(\dfrac{1-\alpha}{2T}\right)t}{\pi t} + \frac{\cos 2\pi\left(\dfrac{1+\alpha}{2T}\right)t}{\dfrac{\pi t}{4\alpha}}\right] \tag{2.111}$$

which is plotted in Figure 2.34. We see that the pulse shape is noncausal, so in practice it cannot be used. However, by truncating the pulse shape to a few symbol periods and delaying it, it is possible to use an approximate version of these root-raised cosine pulse shapes. However, by truncating the pulse shape the PSD of the transmitted signal is no longer as ideally shown in Figure 2.33 because the trunca-

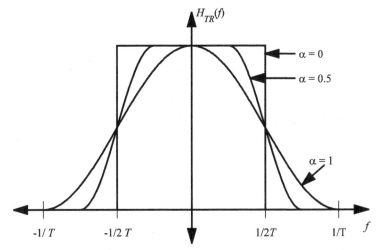

Figure 2.33 Raised cosine shape of the frequency response of the combined transmit and receive filters.

tion will introduce some sidelobes. When α is small, $h_T(t)$ decays more slowly, so truncation has a bigger impact causing these sidelobes to be more pronounced, so typically $\alpha \geq 0.2$ is used in practice.

Although the bandwidth of the transmitted signal is an important characteristic of a modulation scheme, it is also important to use the spectrum efficiently. The spectral efficiency of a modulation scheme is defined as the number of bits that can be transmitted per unit time per unit of frequency, normally measured in bits/second/Hz. It is given by

$$\eta = \frac{\log_2 M}{TB} \tag{2.112}$$

where $\log_2 M$ is the number of bits transmitted per symbol, T is the symbol period (in seconds/symbol), and B is the signal bandwidth (in hertz). For PSK, ASK, and QAM the signal bandwidth depends only on the pulse shape, $h_T(t)$. With a rectangular

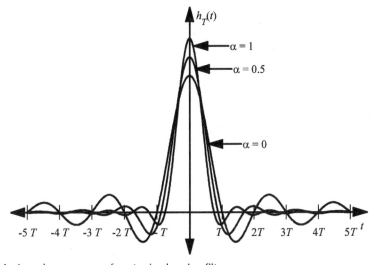

Figure 2.34 Impulse response of root raised cosine filter.

pulse shape, the null-to-null bandwidth of the modulated signal is $2/T$, whereas the bandwidth is $(1 + \alpha)/T$ when a root raised cosine pulse shape with roll-off factor α is used. With a root raised cosine pulse, the spectral efficiency is

$$\eta = \frac{\log_2 M}{1 + \alpha} \qquad (2.113)$$

which increases as M increases. However, with FSK the bandwidth also depends on the modulation order, M. The null-to-null bandwidth of M-ary FSK signal with a rectangular pulse shape and a frequency separation of $f_\Delta = 1/T$ is $(M + 1)/T$, and even with a root raised cosine pulse it is $(M + \alpha)/T$ so the spectral efficiency is

$$\eta = \frac{\log_2 M}{M + \alpha} \qquad (2.114)$$

which decreases quickly as M increases. One may be tempted to decrease the frequency separation in an attempt to improve the spectral efficiency, but this is futile. If f_Δ is not a multiple of $1/2T$, then the transmitted signals are not orthogonal so the signal points become close together, driving up the BER. Even when $f_\Delta = 1/2T$, reliable detection of the signal requires perfect carrier synchronization between the transmitter and receiver, which is not always possible. Furthermore, the spectral sidelobes of the signal decay very slowly when $f_\Delta = 1/2T$, so the effective bandwidth is not much better than when the separation is $1/T$. For these reasons the minimum useable separation is $1/T$.

Example 2.9:   Spectral Efficiency of 4-QAM and 4-FSK

Find the spectral efficiency of 4-QAM and 4-FSK when used with a root raised cosine pulse shape with a roll-off factor of $\alpha = 0.25$.

Solution:
Because $M = 4$ and the bandwidth is $B = (1 + 0.25)/T_x$, for 4-QAM, the spectral efficiency is $\eta = \log_2 4/(1 + 0.25) = 2/1.25 = 1.6$ bits/second/Hz, whereas the spectral efficiency of 4-FSK is $\eta = \log_2 4/(4 + 0.25) = 2/4.25 = 0.47$ bits/second/Hz. Clearly FSK is not a spectrally efficient modulation scheme.

When choosing the pulse shape, it is also important to pay attention to its effect on the *peak-to-average power ratio* (PAPR). The PAPR measures how much the instantaneous signal power can deviate from its average at any given time. Signals with a high PAPR will suffer from greater distortion due to nonlinearities in the RF components of the communication system. We can determine upper bounds on the PAPR for most modulation schemes quite easily. For the transmitted bandpass signal, $v_c(t)$, the instantaneous signal power is $P_v(t) = v_c^2(t)$ which changes very quickly over time as the carrier cycles, and more slowly as the symbols change. The average power is

$$P_{\text{ave}} = \frac{1}{T} \int_0^T \mathrm{E}[v_c^2(t)] \, dt = \frac{E_s}{T} \qquad (2.115)$$

where E_s is the average energy per symbol and is easily determined from the signal constellation. To find the peak instantaneous power, we use the complex lowpass equivalent representation of the bandpass signal. Because $v_c(t) = \Re\{v(t)\sqrt{2}e^{j2\pi f_c t}\}$ from (2.33), we have

$$P_{\text{peak}} = \max_t v_c^2(t) = 2\max_t |v(t)|^2 \qquad (2.116)$$

provided that the carrier frequency is much larger than the bandwidth of $v(t)$. Expressing $v(t)$ in terms of a pulse train and noting that $|a + b| \le |a| + |b|$ gives

$$P_{\text{peak}} = 2\max_t \left|\sum_{n=-\infty}^{\infty} v_n h_T(t - nT)\right|^2 \le 2\max_t \left[\sum_{n=-\infty}^{\infty} |v_n||h_T(t - nT)|\right]^2 \qquad (2.117)$$

We can see that the peak power depends on the sequence of transmitted symbols and on the pulse shape. The summation in (2.117) is maximized when $|v_n|$ is at its maximum value for all n. Let $E_{\max} = \max |v_n|^2$ be the maximum energy of any signal in the constellation. Replacing $|v_n|$ with its maximum value, $\sqrt{E_{\max}}$, gives

$$P_{\text{peak}} \le 2E_{\max} \max_t \left[\sum_{n=-\infty}^{\infty} |h_T(t - nT)|\right]^2 \qquad (2.118)$$

The peak-to-average power ratio is then

$$\text{PAPR} = \frac{P_{\text{peak}}}{P_{\text{ave}}} = 2\frac{E_{\max}}{E_s}\left(T\max_t\left[\sum_{n=-\infty}^{\infty}|h_t(t - nT)|\right]^2\right) = 2\frac{E_{\max}}{E_s}\beta \qquad (2.119)$$

and is usually expressed in decibels. The ratio $E_{\max}/E_s$ depends only on the signal constellation, not on the pulse shape, and formulas to calculate it for various modulation schemes are provided in Table 2.1. The factor of 2 is because we are considering the PAPR of a modulated bandpass signal. This factor would be missing from the PAPR of the corresponding baseband signal. The factor

$$\beta = T\max_t\left[\sum_{n=-\infty}^{\infty}|h_T(t - nT)|\right]^2 \qquad (2.120)$$

is the excess roll-off factor due to pulse shaping and depends only on the pulse shape, not the signal constellation. Maximization to find β needs to be performed over only one symbol period, $t \in [0,T)$, because the summation is periodic. For the rectangular pulse shape, only one term in the summation is nonzero, so $\beta = 1$. For other pulse shapes it may be necessary to use numerical methods to find the peak power. For the root raised cosine pulse, β (in decibels) as a function of the roll-off factor, α, is shown in Figure 2.35. As can be seen, this pulse shape increases the PAPR by between 3.5 and 8 dB.

Table 2.1   Maximum Versus Average Symbol Energy

	PSK, FSK	ASK	QAM
$\dfrac{E_{\max}}{E_s}$	1	$3\dfrac{M-1}{M-\frac{1}{2}}$	$3\dfrac{\sqrt{M}-1}{\sqrt{M}+1}$

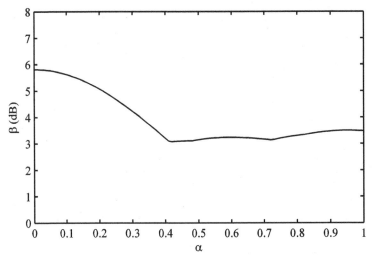

Figure 2.35 Excess PAPR (in decibels) due to a root raised cosine pulse shape with a roll-off factor of α.

Example 2.10:   PAPR of 16-QAM

Find the PAPR of 16-QAM when a root raised cosine pulse shape is used with $\alpha = 0.3$.

Solution:
From Table 2.1 the ratio $E_{\max}/E_s$ for 16-QAM is

$$\frac{E_{\max}}{E_s} = 3\frac{\sqrt{M}-1}{\sqrt{M}+1} = 3\frac{4-1}{4+1} = 1.8 = 2.6 \text{ dB} \tag{2.121}$$

and from Figure 2.35 the excess PAPR when $\alpha = 0.3$ is $\beta = 4.7$ dB. The PAPR is therefore

$$\text{PAPR} = 3 + 2.6 + 4.7 = 10.3 \text{ dB} \tag{2.122}$$

which is 4.7 dB higher than would occur if a rectangular pulse shape were used.

There are also many other modulation schemes that can be used to provide better spectral characteristics. *Continuous phase modulation* (CPM) is a broad class of modulation schemes where the phase of the carrier is prevented from changing abruptly between symbol intervals. By preventing these abrupt changes in the carrier phase, the sidelobes of the PSD are much lower, as is the PAPR. One CPM scheme, *minimum shift keying* (MSK), is a form of continuous-phase frequency shift keying (CPFSK) where the carrier separation is exactly $1/2T$ and the carrier phase changes continuously. The transmitted bandpass signal during the nth symbol interval $(nT \le t < nT + T)$ is given by

$$v_c(t) = A_c \cos(2\pi f_c t + \theta_n + \phi_n(t - nT)/T) \tag{2.123}$$

where $\phi_n \varepsilon \{\pm\pi/2\}$ represents the value of the nth transmitted bit and

$$\theta_n = \sum_{k=0}^{n-1} \phi_k \tag{2.124}$$

depends on all the previously transmitted bits. As such, the transmitted signal at any given time depends not only on the current bit, but all previous ones as well, so this is an example of a modulation scheme with memory. Because the transitions between symbols are continuous and the amplitude is constant, the PAPR is one.

A closely related modulation scheme with good spectral properties that is easy to implement is *offset quaternary phase shift keying* (OQPSK), which is achieved by transmitting two BPSK signals, one on the in-phase carrier and the other on the quadrature phase carrier, just like with 4-QAM, except that the signal on the quadrature phase carrier is delayed by half a symbol period. This means that the *I* and *Q* data never change at the same time. Thus, instead of getting a maximum possible instantaneous phase shift of 180°, the maximum instantaneous phase shift is only 90°. This causes much lower sidelobes in the PSD than with QPSK, although not as low as with MSK. It also leads to a lower PAPR.

Another variant of MSK is Gaussian MSK (GMSK), which involves prefiltering the transmitted data with a Gaussian pulse shape that spans several symbol periods and transmitting the result using frequency modulation. Unlike MSK, where the phase changes continuously but the frequency can change instantaneously, both the phase and frequency change continuously with GMSK. This leads to sidelobes in the PSD that are even lower than MSK, while still providing constant signal envelope. However, the use of the Gaussian pulse shape does lead to the introduction of ISI, which requires some form of equalization at the receiver for compensation. GMSK is used in cellular networks based on the Global System for Mobile Communication (GSM) standard.

2.6.3 Frequency Division Multiple Access

In general, digital communication signals have a bandwidth that is proportional to the inverse of the signal duration; the shorter the signal duration (i.e., the faster the transmission rate), the higher the signal bandwidth, and vice versa. Because the practical signal bandwidth is essentially limited, it is possible for multiple users to simultaneously share the wireless channel, using a technique known as *frequency division multiple access* (FDMA). Each user in a given geographical area is assigned a distinct frequency band in which they can transmit their signals. Provided that the frequency bands do not overlap and each user limits their signal to their assigned frequency band, a receiver can extract the desired signal and block the undesired ones. This concept is illustrated in Figure 2.36, which shows three broadcast television

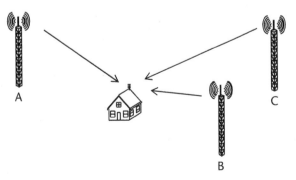

Figure 2.36 FDMA as used for television broadcasting.

signals being received at one house. Each TV station is assigned a different frequency band (channel), as shown in Figure 2.37. By tuning the receiver at the house to one of the frequency bands, the desired signal can be recovered and displayed on the TV screen, without interference from the other signals. Techniques for recovering the desired signal are described in detail in later chapters.

To ensure FDMA works as intended, it is important that the signals do not overlap in the frequency domain. This requires that the transmitted signals have a finite bandwidth. In practice, this is impossible to achieve, as all time-limited signals have infinite bandwidth. As such, all transmitted signals will provide some radiated signal power in frequency bands outside of the assigned frequency band, which will result in interference to the signals in the other frequency bands. Because this out-of-band signal power is typically strongest in the immediately neighboring frequency bands, this type of interference is referred to as *adjacent channel interference* (ACI). When a receiver is trying to detect a weak signal from a distant transmitter, while a nearby transmitter is transmitting a strong signal in an adjacent frequency band, strong ACI can overlap the desired signal, and there is very little that the receiver can do to mitigate its effects. It is therefore necessary to carefully design transmitters so they limit the amount of power transmitted outside of the allocated frequency band. This can be done by careful selection of the pulse-shaping filter and avoiding unwanted nonlinear effects in the RF components of the transmitter. Guard bands (unallocated frequency bands) can also be placed between the assigned bands to reduce the effects of ACI.

A related issue with FDMA is *cochannel interference* (CCI). Because the available frequency spectrum is limited, it is desirable to reuse it in different geographical areas. For example, the same frequency band can be assigned to two different TV stations, provided the two stations are sufficiently far apart, such as in different cities. CCI arises when the signal from a distant transmitter arrives at a sufficiently high power level to interfere with the signal from a nearby transmitter using the same assigned frequency band. In many cases, the assignment of frequency bands is performed by central governments whose regulatory bodies (such as the FCC in the United States) ensure that CCI is not a serious problem. However, some frequency bands are assigned to independent organizations, such as cellular telephone network carriers, who subdivide the bands into smaller subchannels that are assigned to individual users. To increase the capacity of their networks, the carriers want to make the same subchannels available to as many users as possible in different areas, by decreasing the size of the area where each subchannel is used. As such, the carriers

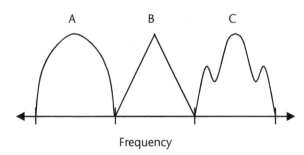

Frequency

Figure 2.37   Frequency allocation for different TV stations with FDMA.

spend considerable effort balancing system capacity (in terms of the number of supported users) with the adverse effects of CCI. In the unlicensed frequency bands used by home wireless LANs (such as those based on the IEEE 802.11 standards), CCI can be more problematic because there is no centralized authority allocating channels to users.

2.7 Wireless Channel Models

Although the AWGN channel model presented earlier in this chapter is very important, it only reflects a small aspect of the issues facing wireless communication. In reality, the communication channel is much more complicated. At the very least, there will be signal attenuation and propagation delay, and there will likely also be multipath interference. The causes and implications of signal distortion in wireless channels are discussed in this section.

2.7.1 Signal Attenuation

In wireless communication systems the transmitted signal is usually severely attenuated, so the received signal is often extremely weak. Some factors that affect the attenuation include the antenna gains, the distance-dependent path loss, and shadowing due to obstructions between the transmitter and receiver.

Antennas are needed in wireless systems to transform between RF signals in the electronic circuits and electromagnetic (EM) signals in the air. The antenna attached to the transmitter radiates the EM signal into the air, and the receiving antenna collects energy from the air and provides it to the receiver so that the signal can be detected and processed. As with any conversion process there is always an associated loss. That is, not all of the RF signal energy is converted to EM energy and vice versa. We can define the antenna efficiency as the ratio of these two energies.

Antennas can either be omnidirectional (also called isotropic), which means that energy radiates equally in all directions, or directional with more energy focused in a particular direction. Similarly, a directional receive antenna collects more energy arriving from a particular direction and less from other directions. The directivity of an antenna coupled with its efficiency can be described by the *directional antenna gain*, $G_A(\theta, \phi)$, where θ, ϕ define the direction in spherical coordinates. The antenna gain refers to the amount of excess energy radiated in a given direction over and above that which would have been radiated in the same direction by an isotropic antenna, and is usually expressed in units of decibels-isotropic (dBi). Although the radiation pattern is three-dimensional, it is common to plot it using two-dimensional cross-sectional views, such as shown in Figure 2.38 which shows an example of the radiation pattern along the horizontal plane. The distance between the origin and any point on the curve is proportional to the antenna gain in the direction of the point, so in this example the bulk of the energy is radiated in the main beam along the positive x-axis.

Note that the antenna gain is often quoted as its maximum value, $G_A = \max G_A(\theta, \phi)$, along with the width of the main beam (in degrees). The 3-dB beamwidth is often used to specify the range of directions for which the gain is within 3 dB of its

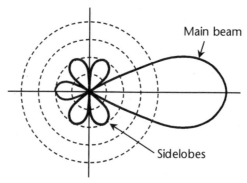

Figure 2.38 Example of an antenna radiation pattern.

maximum. Generally, the more directional an antenna is, the greater the maximum gain is. Because the total radiated energy does not depend on the antenna, when less energy is radiated in unwanted directions more can be radiated in the desired direction.

The energy radiated from the transmit antenna becomes more diffuse the farther it travels. The receive antenna will only capture a small fraction of the transmitted energy, so if a transmit antenna radiates a signal in all directions through free space, the power at the receive antenna will be

$$P_R = P_T \frac{A_R}{4\pi d^2} \tag{2.125}$$

where P_T is the transmitted power, A_R is the effective cross-sectional area of the receive antenna, d is the distance between the transmitter and receiver, and $4\pi d^2$ is the surface area of a sphere with radius d. As d increases a smaller fraction of the surface area of the sphere will pass through an area of A_R, so less of the transmitted energy can be collected. The effective area of the receive antenna is related to the antenna gain in the direction of the received signal, and is given by

$$A_R = G_R \frac{\lambda_c^2}{4\pi} \tag{2.126}$$

where $\lambda_c = c/f_c$ is the wavelength of the signal, c is the speed of light, f_c is the carrier frequency, and G_R is the gain of the receive antenna. Isotropic antennas, which receive power from all directions, have a gain of $G_R = 1$, but it is possible to use directional antennas focused in a narrow beam. Also taking into account the gain of a directional antenna at the transmitter, with a gain of G_T in the direction of the receiver, we can therefore express the received signal power as

$$P_R = P_T \frac{G_T G_R}{(4\pi d / \lambda_c)^2} \tag{2.127}$$

or, in decibels,

$$P_{R\|dB} = P_{T\|dB} + G_{T\|dB} + G_{R\|dB} - L_{P\|dB}(d) \tag{2.128}$$

where $L_{P\|dB}(d) = 10\log_{10}(4\pi d / \lambda_c)^2$ is the free-space path loss at distance d.

In practice, according to experimental measurements, the received signal power decays more quickly than the square of the distance, and there is considerable variation between measurements taken at different locations at the same distance from the transmitter, due to shadowing effects from obstructions, such as buildings, trees, and rainfall, in the transmission path. A better model for the received signal power is

$$P_{R\|dB} = P_{T\|dB} + G_{T\|dB} + G_{R\|dB} - L_{P\|dB}(d_0) - 10\rho\log_{10}\frac{d}{d_0} - L_{\sigma\|dB} \qquad (2.129)$$

where d_0 is the close-in reference distance (a distance close to the transmit antenna but still in the far field of the antenna), ρ is the path-loss exponent that measures how quickly the path loss increases with distance, and $L_{\sigma\|dB}$ represents the shadowing. The path loss at the close-in reference distance is either calculated according to $L_{P\|dB}(d_0) = 10\log_{10}(4\pi d/\lambda_c)^2$ or is taken from measurements. The path-loss exponent is usually in the range from 2 to 6 depending on the environment. Typical values are shown in Table 2.2. The attenuation due to shadowing, $L_{\sigma\|dB}$, typically follows a Gaussian distribution with a standard deviation of σ_S (in decibels). In a linear scale, L_σ has a log-normal distribution, so it is usually referred to as log-normal shadowing.

Example 2.11:   Received Signal Power

Suppose a transmitter with a transmit antenna gain of $G_{T\|dB} = 8$ transmits a signal with a power of 20 dBm and a carrier frequency of 2.4 GHz to a receiver that is 100 meters away. The receive antenna has a gain of $G_{R\|dB} = 6$. Find the received signal power, using a close-in reference distance of $d_0 = 1$ meter and a path-loss exponent of $\rho = 3.5$. Ignore the log-normal shadowing.

Solution:
The wavelength of the signal is $\lambda_c = 3 \times 10^8/2.4 \times 10^9 = 0.125$, so the path loss at distance d_0 is $L_{P\|dB}(d_0) = 10\log_{10}(4\pi/0.125)^2 = 20$ dB. The received signal power is

$$P_{R\|dBm} = 20 + 8 + 6 - 20 - 10 \times 3.5 \times \log_{10}\frac{100}{1}$$

$$= 14 - 70 = -56 \text{ dBm} \qquad (2.130)$$

which is 76 dB less than the transmitted power.

Because the received signal is usually extremely weak, it is necessary to use amplifiers at the receiver to boost the signal level to a usable level. Doing so, however,

Table 2.2   Path-Loss Exponent in Various Environments

Environment	*Path-Loss Exponent, ρ*
Free space	2
Urban cellular radio	2.7 to 3.5
Shadowed urban cellular radio	3 to 5
In building line of sight	1.6 to 1.8
Obstructed in buildings	4 to 6
Obstructed in factories	2 to 3

increases the amount of noise in the received signal. Furthermore, because the wireless communication environment is constantly changing, the attenuation due to path loss and shadowing is typically time-variant. Some modulation schemes, such as ASK and QAM, require knowledge of the received signal strength to make reliable decisions, so the time-varying attenuation must be estimated, tracked, and compensated for. However, the decision rules for PSK and FSK do not depend on the signal strength, allowing for less complex receiver design in this regard.

2.7.2 Propagation Delay

Because radio waves only travel at the speed of light through free space, there will be a transmission delay corresponding to time it takes for the signal to reach the receiver. Although this delay may be only a few nanoseconds, because it is time-varying as the user moves around, the delay causes synchronization issues that need to be addressed. A channel model incorporating the effects of attenuation and propagation delay is

$$r_c(t) = \alpha v_c(t - \tau) + w_c(t) \tag{2.131}$$

where α is the signal attenuation and τ is the propagation delay, while $r_c(t)$ is the received bandpass signal, $v_c(t)$ is the transmitted signal, and $w_c(t)$ is the additive white Gaussian noise.

To illustrate the effects of the propagation delay, consider the complex lowpass equivalent representation of the transmitted signal based on (2.15). The delayed transmitted signal is

$$\begin{aligned} v_c(t - \tau) &= \Re\{v(t - \tau)\sqrt{2}e^{j2\pi f_c(t-\tau)}\} \\ &= \Re\{v(t - \tau)\sqrt{2}e^{j2\pi f_c t}e^{-j2\pi f_c\tau}\} \\ &= \Re\{v(t - \tau)\sqrt{2}e^{j2\pi f_c t}e^{j\phi_c}\} \end{aligned} \tag{2.132}$$

We can see that the propagation delay causes a corresponding delay in the complex lowpass equivalent signal, $v(t)$, and a phase rotation of the carrier by $\phi_c = j2\pi f_c\tau$. Both these artifacts of the propagation delay are addressed separately, by clock recovery and carrier synchronization techniques.

2.7.2.1 Clock Recovery

Uncertainty in the delay requires clock recovery techniques to determine the optimal times to sample the received lowpass signal. From (2.40), the received lowpass signal, delayed by τ seconds, is given by

$$r(t) = \sum_{n=0}^{N_s - 1} v_n h_{TR}(t - nT - \tau) \tag{2.133}$$

where the additive noise and the carrier phase rotation have been neglected to focus the discussion on the issues most relevant to clock recovery. We can express the propagation delay as the sum of a large-scale delay (the number of full symbol periods in the delay) and a small-scale delay (the fractional delay within one symbol period), so $\tau = k_o T + \tau_o$ where k_o is the number of full symbols, and $\tau_o \in [-T/2, T/2]$ is the fractional delay. If no attempt were made at clock recovery, the received samples according to (2.41) and (2.133) would be

$$r_n = r([n+1]T) = \sum_{m=0}^{N_s-1} v_m h_{TR}([n-m+1]T - k_o T - \tau_o) \qquad (2.134)$$

The large- and small-scale delays have different effects on the received samples, and are handled using different techniques at the receiver.

If τ_o is zero (or known and compensated for), and ISI-free pulse shaping is used so (2.42) is valid, the received samples are

$$r_n = \sum_{m=0}^{N_s-1} v_m h_{TR}([n-m-k_o+1]T) = v_{n-k_o} \qquad (2.135)$$

That is, the received samples are equal to the transmitted symbols, but delayed by k_o samples. The receiver must have knowledge k_o to determine when the first symbol in the packet arrives so that prior samples can be discarded. This is accomplished by frame synchronization. One simple method is to transmit a known training sequence prior to transmission of the data symbols. With knowledge of the training sequence the receiver can observe the output of the decision device until the training sequence is found, and the following sample would be for the first data symbol.

If k_o is known and compensated for by delaying the sampling by $k_o T$ seconds, the received samples are

$$r_n = r([n+1]T + k_o T) = \sum_{m=0}^{N_s-1} v_m h_{TR}([n-m+1]T - \tau_o) \qquad (2.136)$$

If τ_o is nonzero, then ISI will occur, even if the pulse shape is nominally ISI-free, as (2.42) is only valid if the signal is sampled at the optimal times (that is, if the collection of the samples is delayed by τ_o seconds). The ISI will result in a very high BER, and although an adaptive equalizer can help combat the ISI if it is not too severe, there may still be some reduction in performance, so it is usually preferred to estimate τ_o and delay the sampling accordingly.

An analog clock recovery circuit, such as an early-late gate synchronizer, can be used to dynamically delay or advance the sampling times slightly in an attempt to lock onto the optimal times. That is, the samples are collected with a delay of $\hat{\tau}_o$, and $\hat{\tau}_o$ is dynamically adjusted over time until $\hat{\tau}_o = \tau_o$. Digital clock recovery techniques typically involve collecting multiple samples per symbol and analyzing several symbols-worth of these samples in an attempt to determine τ_o. This analysis can be aided with the use of a training sequence. With a training sequence the receiver knows what the samples should be if they were collected at the correct times. An estimate of the fractional delay, $\hat{\tau}_o$, can be determined as that value of the fractional delay that would give received samples closest to their expected values. The received samples with delay $\hat{\tau}_o$ can then be estimated by interpolating between the collected samples and these samples can be passed on to the decision device. The analog sampling device can also be adjusted to sample with a delay of $\hat{\tau}_o$ so that future samples are collected closer to the optimal times. The initial estimate of τ_o can be further refined by using feedback of the decisions made by the decision device, allowing for the receiver to track changes in the propagation delay and to handle small mismatches between the clock frequencies at the transmitter and receiver. It is important to note that frame synchronization and clock recovery with training sequences should be carried out jointly, because frame synchronization requires detection of the training sequence, which can only be achieved if the correct fractional delay has been determined, which, in turn, requires knowledge of when the training sequence begins.

2.7.2.2   Carrier Synchronization

When information is carried in the phase of the carrier wave, such as with PSK or QAM, it is important for the receiver to maintain carrier synchronization with the transmitter. The received signal is demodulated by mixing it with a locally generated carrier reference signal, which should have the same phase as the received carrier. The phase of the received carrier depends on the reference phase of the transmitted carrier, and the phase rotation due to the propagation delay, ϕ_c. If the received signal, with carrier reference phase ϕ_c, is mixed with a carrier reference signal with phase $\hat{\phi}_c$, the downconverted signal from (2.36) is

$$r_0(t) = \Re\left\{ v(t)\sqrt{2}e^{j2\pi f_c t + j\phi_c} \right\} \sqrt{2}e^{-j2\pi f_c t - j\hat{\phi}_c} = v(t)e^{j(\phi_c - \hat{\phi}_c)} \tag{2.137}$$

where the additive noise and the high-frequency replica of the signal have been ignored for simplicity. If we define $\phi_\in = \phi_c - \hat{\phi}_c$ as the carrier phase mismatch, the received samples are

$$r_n = v_n e^{j\phi_\varepsilon} \tag{2.138}$$

That is, all the received samples are rotated by ϕ_ε. This phase rotation will cause the BER to be much worse. A scatter plot of the received samples for a 4-QAM signal is shown in Figure 2.39 when there is a phase rotation of 30°. As can be seen, many samples fall near the boundaries of the decision regions, which are the horizontal and vertical axes. These samples are likely to be detected incorrectly. If the phase rotation is greater than ±45°, then the majority of samples will not be detected correctly, even if there is no noise. One technique to avoid this problem is to ensure that the carrier reference signal has a phase of $\hat{\phi}_c = \phi_c$ so that $\phi_\varepsilon = 0$. This is known as *coherent demodulation*, and can be accomplished by using a carrier recovery circuit such as a phase locked loop, as discussed in Chapter 6, to generate the carrier reference signal.

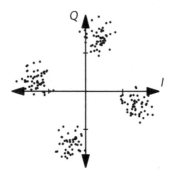

Figure 2.39   Received 4-QAM samples with a 30° phase rotation.

Building a circuit that recovers the absolute phase of the received signal is difficult, so *noncoherent demodulation* is often used as a low-cost alternative. With noncoherent demodulation the received signal is demodulated with a carrier reference signal with arbitrary phase, so all the received samples are rotated by ϕ_ε, which can take on any value over $[0, 2\pi)$. This phase rotation is the same for all the received samples in a packet, or varies fairly slowly as ϕ_c changes. Digital signal processing, perhaps aided by a known training sequence or occasional pilot symbols, can be used to estimate ϕ_ε, and the received samples can be rotated back to their correct positions.

Alternatively, modulation schemes that do not require knowledge of the carrier phase, such as ASK and FSK, can be used. With ASK the receiver only needs to know the amplitude of the carrier, and with FSK the receiver needs the amplitudes of all the carriers and selects the largest. With both these schemes, if noncoherent demodulation is used, then demodulation must be performed using both in-phase and quadrature phase carriers, even though the data is only transmitted on the in-phase carrier, because the phase rotation may move a significant portion of the transmitted signal onto the quadrature phase carrier. Because the noise on both the in-phase and quadrature phases affects the decision, noncoherently detected ASK and FSK have a higher BER (or require 3 dB more signal power) than when detected coherently, but have a more simple receiver design without the carrier recovery circuit. Furthermore, noncoherently detected FSK requires a carrier separation of f_Δ that is a multiple of $1/T$ to ensure orthogonality in both the in-phase and quadrature phase carriers, versus the minimum separation of only $1/2T$ required for coherent detection.

It is also possible to use a version of PSK that does not need any knowledge of the absolute phase of the transmitted signal. This type of modulation is called *differential phase shift keying* (DPSK). With DPSK the information is transmitted not directly in the phase of the carrier as with PSK, but as the change in the carrier phase over two consecutive symbols. The receiver compares the phase of current sample to the phase of the previous sample, and the change in phase tells the receiver what symbol was transmitted. Thus, for instance, in binary DPSK if there was no change in phase, a zero was transmitted and if the phase changes by 180°, then a one was transmitted. Because the decision on the current symbol is affected by noise in both the current symbol and the previous one, DPSK suffers from a higher BER than coherently detected PSK.

2.7.3   Multipath Interference and Fading

As mentioned, signals transmitted wirelessly travel in all directions from the antenna. They are then reflected, diffracted, and scattered by objects in the environment, and some of these replicas of the transmitted signal will inevitably arrive at the receiver. As such, there are multiple paths between the transmitter and the receiver, as shown in Figure 2.40, for example. Each of these paths has its own attenuation and delay, and the received signal is the composite of the signals along each of the paths, so

$$r_c(t) = \sum_k \alpha_k v_c(t - \tau_k) + w_c(t) \tag{2.139}$$

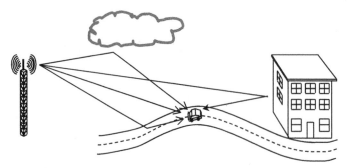

Figure 2.40 Multipath interference in wireless communication systems.

where α_k and τ_k are the attenuation and propagation delay along the kth path, and the summation is over all paths.

Because $v_c(t)$ is essentially a sine wave, signals along two paths that arrive at nearly the same time may combine constructively or destructively. Making use of the complex lowpass equivalent signal notation we have

$$r_c(t) = \sum_k \alpha_k \Re\{v(t - \tau_k)\sqrt{2}e^{j2\pi f_c(t - \tau_k)} + w_c(t)$$

$$= \Re\left\{\sum_k \alpha_k e^{j\phi_k} v(t - \tau_k)\sqrt{2}e^{j2\pi f_c t}\right\} + w_c(t) \tag{2.140}$$

where $\phi_k = -2\pi f_c \tau_k$, If all the paths arrive at nearly the same time, so that $\tau_k \cong \tau_0$, we can write (2.140) as

$$r_c(t) = \Re\left\{\left(\sum_k \alpha_k e^{j\phi_k}\right)v(t - \tau_0)\sqrt{2}e^{j2\pi f_c t}\right\} + w_c(t)$$

$$= \Re\{(\alpha e^{j\phi})v(t - \tau_0)\sqrt{2}e^{j2\pi f_c t}\} + w_c(t) \tag{2.141}$$

Depending on the phases, $\{\phi_k\}$, the terms in the summation

$$\alpha e^{j\phi} = \sum_k \alpha_k e^{j\phi_k} \tag{2.142}$$

may cancel each other out, causing the received signal to be very weak. Although the path attenuations $\{\alpha_R\}$ in (2.142) change over time, they do so quite slowly. However, the phases can change very quickly because movement of less than half a wavelength can cause a change in the propagation delay by enough to result in a 180° change in the phase. As a result, motion of only a few centimeters can cause $\alpha e^{j\phi}$ to change significantly. As a result, over time the received signal power changes quickly about its mean due to multipath fading. The mean power itself also changes over time because of changes to the path loss and shadowing, but this occurs much more slowly.

Figure 2.41 shows an example of how the instantaneous and average receive power varies over time. Although the average power changes slowly over time due to changes in the path loss and shadowing, multipath interference causes the instantaneous power to change about the average very quickly. This phenomenon is referred to as *multipath fading* or *Rayleigh fading*. The instantaneous power often fades by up to 20 dB or more below the average.

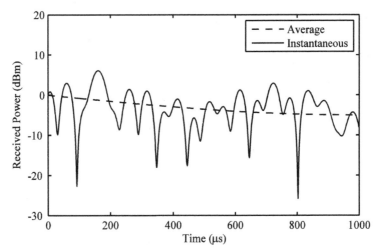

Figure 2.41 Variation of instantaneous received power due to multipath fading.

When the number of paths is large, the law of large numbers suggests that $\alpha e^{j\phi}$ can be well approximated by a complex Gaussian random variable. It has a phase that is uniformly distributed over $[0,2\pi)$ and an amplitude that has a Rayleigh distribution with a probability density function of

$$f_\alpha(\alpha) = \frac{2\alpha}{\Omega} e^{-\alpha^2/\Omega} \text{ for } \alpha \geq 0 \tag{2.143}$$

where the parameter $\Omega = E[\alpha^2]$ is the average power attenuation of the channel due to other factors such as path loss and shadowing as provided by (2.129).

The instantaneous received SNR is given by

$$\gamma_b = \alpha^2 \frac{E_b}{N_0} \tag{2.144}$$

where E_b is the transmitted energy per bit and α^2 is the instantaneous power attenuation. Because α is random, with a Rayleigh distribution, γ_b is also random, but with a chi-squared distribution, so its pdf is

$$f_\gamma(\gamma_b) = \frac{1}{\overline{\gamma_b}} e^{-\gamma_b/\overline{\gamma_b}} \tag{2.145}$$

where

$$\overline{\gamma_b} = E\{\gamma_b\} = E[\alpha^2]\frac{E_b}{N_0} = \Omega\frac{E_b}{N_0} \tag{2.146}$$

is the average received SNR, which depends on the attenuation due to pathloss and shadowing but not on fading.

Multipath fading causes a serious impediment to reliable wireless communication. In the absence of fading, the BER depends on the instantaneous SNR. In a fading environment, where the instantaneous SNR is random, we are more interested in the average BER, averaged over the instantaneous SNR. The average BER is given by

$$P_b = E[P_b(\gamma_b)] = \int_0^\infty P_b(\gamma_b) f_\gamma(\gamma_b) \, d\gamma_b \qquad (2.147)$$

where $P_b(\gamma_b)$ is the BER, which depends on the modulation scheme, when the instantaneous SNR is γ_b. For example, when BPSK is used,

$$P_b(\gamma_b) = \frac{1}{2} \text{erfc}\left(\sqrt{\gamma_b}\right) \qquad (2.148)$$

Substituting (2.148) for $P_b(\gamma_b)$ and (2.145) for $f_b(\gamma_b)$ in (2.147), and performing the integration yields

$$P_b = \frac{1}{2}\left[1 - \sqrt{\frac{\overline{\gamma_b}}{1 + \overline{\gamma_b}}}\right] \qquad (2.149)$$

The average BER of BPSK in a Rayleigh fading channel as a function of $\overline{\gamma_b}$ is shown in Figure 2.42, along with the BER in an AWGN channel without fading for comparison. As can be seen, to achieve a target BER, the average SNR must be much larger for fading channels than for the AWGN channel. In fact, the BER drops by only one order of magnitude for each 10 dB increase in the SNR.

In the previous discussion, we considered the case when the signals along all the paths arrive at essentially the same time. In practice, there is usually some dispersion in the propagation delays. The *delay spread* of the channel is the time difference between the shortest and longest delays. If the delay spread is longer than a fraction of the symbol duration, then each transmitted symbol will be spread over more than one symbol interval, resulting in intersymbol interference. As a result, the channel no longer has a flat frequency response, as some frequencies are attenuated more than others. Because the frequency response is also time-variant, this type of multipath interference is referred to as *frequency-selective fading*.

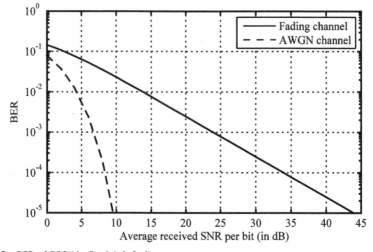

Figure 2.42 BER of BPSK in Rayleigh fading.

2.8 Advanced Communication Techniques

To provide better performance in wireless communication systems, many more advanced communication techniques have been developed and deployed in commercial systems. These techniques, among other things, are designed to help overcome the difficulties caused by multipath fading, both frequency flat and frequency selective. For example, the use of orthogonal frequency division multiplexing allows for easy frequency-domain equalization of frequency-selective fading channels. Using multiple antennas at the transmitter and/or the receiver allows for beamforming, spatial diversity against fading, and spatial multiplexing for higher throughput. Spread spectrum systems allow for better protection against interference and jamming and easy multiuser spectrum sharing. Error control coding allows for the detection and correction of transmission errors, significantly improving the reliability of the communication system. These techniques are described in more detail in this section.

2.8.1 Orthogonal Frequency Division Multiplexing

Orthogonal frequency division multiplexing (OFDM) is a modulation scheme that is designed to improve the performance in radio links that suffer from frequency selective fading. The idea is to replace a single carrier with a wide bandwidth with multiple carriers (called subcarriers) that individually have a narrow bandwidth. Thus, instead of having one carrier with a bandwidth of 8 MHz, for example, there can be eight subcarriers each with a bandwidth of 1 MHz. Current technology typically has the number of subcarriers somewhere between 16 and 1,024.

During each OFDM symbol interval, N different PSK or QAM symbols are transmitted simultaneously, with each symbol transmitted on a separate subcarrier. Let V_k denote the signal constellation point for the symbol transmitted on the kth subcarrier. The transmitted bandpass OFDM signal is the sum of the signals on all the subcarriers, so

$$v_c(t) = \sum_{k=0}^{N-1} \Re \left\{ V_k h_T(t)\sqrt{2}e^{j2\pi f_k t} \right\} \tag{2.150}$$

where $h_T(t)$ is a normalized rectangular pulse of duration T and f_k is the carrier frequency of the kth subcarrier, given by $f_k = f_c + k/T$. The separation between the subcarriers is chosen to be $1/T$, which makes the subcarrier signals orthogonal even though they overlap in frequency. This removes the need for guard bands between subcarriers, allowing for the tightest packing of the subcarriers for a given bandwidth. Figure 2.43 shows the PSD of an OFDM signal using $N = 8$ subcarriers and a nominal carrier frequency of $f_c = 10/T$. The dashed lines are the PSDs of the signals on the individual subcarriers.

To avoid the need for multiple RF signal chains, one for each subcarrier, OFDM signals are generated by using clever digital signal processing. The data to transmit is first passed through a serial-to-parallel converter as shown in Figure 2.44. This breaks the data into N data streams, one for each subcarrier. The data in each stream are mapped to points in the signal constellation, which is typically QAM or PSK. Note that it is not necessary to use the same constellation for each subcarrier. Because each symbol is transmitted on a different subcarrier, we can view these symbols, $\{V_k\}$, as

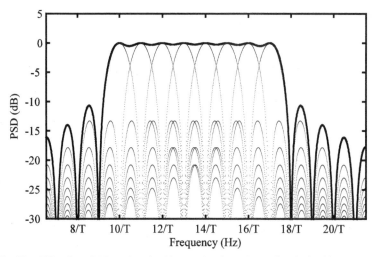

Figure 2.43 The PSD of an OFDM signal with $N = 8$ subcarriers. The dashed lines are the PSDs of the signals on the individual subcarriers.

a frequency-domain representation of the transmitted signal. By applying an inverse fast Fourier transform (IFFT) to the symbols, we can get a time-domain representation of the signal. In fact, the values at the output of the IFFT, $\{v_n\}$, given by

$$v_n = \sum_{k=0}^{N-1} V_k e^{j2\pi\frac{kn}{N}} \tag{2.151}$$

are just samples of the complex lowpass equivalent signal of $v_c(t)$ in (2.150),

$$v(t) = \sum_{k=0}^{N-1} V_k h_T(t) e^{j2\pi kt/T} \tag{2.152}$$

sampled at $t = nT/N$. Therefore, passing $\{v_n\}$ through a digital-to-analog converter (DAC) will give $v(t)$. Before doing so, it is necessary to first insert a *cyclic prefix*. Because OFDM signals are usually intended for transmission over frequency selective fading channels, we know the transmitted signal will be distorted by the convolution with the channel impulse response. This will cause the OFDM signal to spread out over a longer duration than T seconds when received. To prevent consecutive OFDM symbols from interfering with each other, it is a good idea to insert a guard interval between symbols, as shown in Figure 2.45. The duration of the guard interval should be as long as the maximum expected delay spread of the channel. However, instead of not transmitting anything during the guard interval, the last few samples of the following OFDM symbol are transmitted during the guard interval. That is, for $n \, \varepsilon \{1,2,...,L\}$ where L is the delay spread of the channel in samples, we

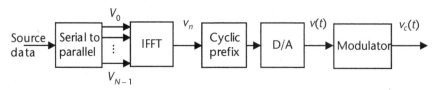

Figure 2.44 Block diagram of an OFDM transmitter.

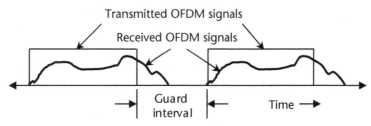

Figure 2.45   Guard intervals between consecutive OFDM signals.

set $v_{-n} = v_{N-n}$. Doing so allows for simple equalization at the receiver to counteract the distortion. The resulting sequence of samples are passed through the DAC and then upconverted by the modulator to the desired frequency band.

After transmission, the received signal is down-converted, filtered, and sampled. An FFT is applied to the received samples to recover the transmitted symbols. Because of the inclusion of the cyclic prefix at the transmitter, the received samples at the output of the FFT are related to the transmitted symbols by

$$R_k = H_k V_k + W_k \tag{2.153}$$

where W_k is an additive noise sample and H_k is the complex channel gain of the kth subcarrier. After dividing R_k by H_k, a decision device can recover the transmitted bits.

If the channel is not frequency-selective, the bit error rate for OFDM is the same as that for the single carrier case using the same modulation scheme, so in straight line-of-sight environments like satellite there is little advantage to using OFDM. The biggest benefits come with frequency selective channels, where there is no need for a complex time-domain equalizer (which is usually implemented with an adaptive tapped delay line), because each subcarrier essentially experiences frequency flat fading. As shown in Figure 2.46, some subcarriers may have severe fading while others are very good. In this example, the symbols transmitted on subcarriers 2 and 5 are severely attenuated and it may not be possible to reliably recover the corresponding transmitted bits. If the transmitter knows the channel frequency response, it can avoid transmitted data on the poor subcarriers. If not, the transmitter can use channel coding so the receiver can recover data transmitted on the weak subcarriers based on the data received on the other subcarriers.

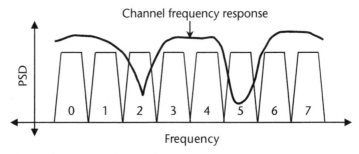

Figure 2.46   Effect of frequency-selective channel on different subcarriers.

OFDM also offers some bandwidth advantages. If a single carrier tone is used to transmit, say, 20 MB/s and a 25% guard band is used then the total signal will occupy 25 MHz of spectrum. If the same data is transmitted using 10 subcarriers, with a 25% guard band on only the first and last subcarriers, then only 20.5 MHz is needed. Remember that the subcarriers do not need guard bands from each other because they are orthogonal. Furthermore, by reducing the power in subcarriers near the edges, and by using an appropriate windowing function prior to the DAC, it is possible to generate signals with very low sidelobes.

OFDM can also be used as part of a multiple-access scheme to transmit different data streams to different receivers (users) simultaneously. This scheme is known as orthogonal frequency division multiple access (OFDM). The set of subcarriers is divided into subsets, and each subset is assigned to a different user. Data for each user is transmitted only on the subcarriers assigned to the user, while data for other users are transmitted on the other subcarriers. This is very similar to regular FDMA, except OFDM subcarriers are used instead of separate carriers. It does, however, have some important advantages over FDMA. Because each user receives the entire OFDM signal, it recovers the data on all the subcarriers and then discards the unwanted data. This means it is very easy for the system to dynamically allocate subcarriers to users without any need to tune the receivers to different frequencies. Furthermore, with OFDMA some users can easily be assigned more subcarriers than others if they have different throughput requirements, whereas in a traditional FDMA system it is difficult to dynamically adjust the signal bandwidth. In addition, because each user experiences different channel conditions, some subcarriers may be severely faded for one user but very strong for another. Careful assignment of subcarriers to users can ensure that each user is served only with good subcarriers, while each subcarrier is likely to be good for at least one user.

Although OFDM provides many advantages over single-carrier systems, the transmitted signal does not have a constant envelope, even when using constant envelope signals with PSK on each subcarrier. In fact, OFDM signals have a notoriously high peak-to-average power ratio (PAPR). The worst-case peak power will occur when the same symbol is transmitted on all subcarriers and this symbol has the largest energy, E_{max}. At $t = 0$ the lowpass signal of (2.152) will have an amplitude of $N\sqrt{E_{max}/T}$, so the peak power of the corresponding bandpass OFDM signal will be

$$P_{peak} = 2N^2 E_{max}/T \tag{2.154}$$

Because the average power of an OFDM signal is $P_{ave} = NE_s/T$, the PAPR is

$$\text{PAPR} = \frac{P_{peak}}{P_{ave}} = 2N \frac{E_{max}}{E_s} \tag{2.155}$$

where the ratio E_{max}/E_s is given in Table 2.1 depending on the modulation scheme. Note that it is extremely unlikely that this peak power will be reached when the number of subcarriers is large, so using a maximum value of $N = 16$ in this equation gives a more realistic PAPR.

Example 2.12:   PAPR of OFDM

Suppose an OFDM system with $N = 64$ subcarriers is transmitting 16-QAM symbols on each subcarrier. Find the PAPR.

Solution:
From Table 2.1 the ratio E_{max}/E_s for 16-QAM is

$$\frac{E_{max}}{E_s} = 3\frac{\sqrt{M}-1}{\sqrt{M}+1} = 3\frac{4-1}{4+1} = 1.8 = 2.6 \text{ dB} \tag{2.156}$$

Using (2.155) with N only equal to 16 (since it is unlikely that more than 16 of the subcarriers will contribute to the peak power), the PAPR is

$$PAPR = 2 \times 16 \times 1.8 = 57.6 = 17.6 \text{ dB} \tag{2.157}$$

which is 7.3 dB higher than for single-carrier 16-QAM as found in Example 2.10.

The high PAPR of OFDM places an extra burden on the power efficiency of the amplifiers in the transmitter, leading to the need for more expensive components. As a result, OFDM is less attractive in the uplink of a cellular system, where it is desirable to keep the cost of the mobile devices as low as possible. Although some signal processing techniques have been proposed that reduce the PAPR of the transmitted signal, these techniques are often quite complex or degrade the system performance. For the uplink it is preferable to use a single carrier system instead of OFDM. However, to allow for easy spectrum sharing like OFDMA, a novel modulation scheme known as *single-carrier FDMA* (SC-FDMA) is used for the uplink in LTE systems.

SC-FDMA allows the transmitter to dynamically adjust the bandwidth and the frequency of its transmitted signal without needing to adjust the RF components. This is accomplished by using a bit of additional signal precoding prior to generating an OFDM-like signal, as shown in the block diagram in Figure 2.47. The source data are mapped to modulation symbols such as QAM or PSK and then converted to the frequency domain using an M-point FFT. The M symbols at the output of the FFT are then applied to M of the inputs of an N-point IFFT, where $N \geq M$, and values of 0 are supplied to the other $N - M$ inputs. The output of the IFFT is processed in a manner identically to the output of the IFFT in an OFDM transmitter. Recalling that the input to the IFFT in an OFDM system is the signals to transmit on each of the subcarriers, we see that each transmitter in an SC-FDMA system uses only a few of the subcarriers. By changing which inputs to the IFFT to which we connect the output of the FFT, we can control which subcarriers are used, and

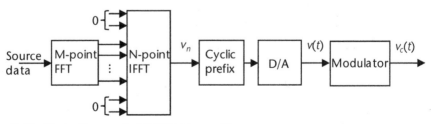

Figure 2.47   Block diagram of a SC-FDMA transmitter.

hence control the carrier frequency of the single-carrier signal. By changing M, we can control the number of subcarriers used, thereby controlling the bandwidth of the transmitted signal.

Although the SC-FDMA transmitter looks very similar to the OFDM transmitter, the inclusion of the FFT changes the nature of transmitted signal. Instead of transmitting modulation symbols on each subcarrier, we are transmitting the frequency domain equivalent of the symbols. The net result is that the time-domain-transmitted signal is, in fact, a single-carrier, not multicarrier, signal. As such, it does not have nearly the same problems with the PAPR.

The receiver for a SC-FDMA system is very similar to that for an OFDM system. After taking the N-point FFT of the received samples, frequency-domain equalization is performed just as with an OFDM system. Then an M-point IFFT is taken of the appropriate outputs of the FFT (from the same subcarriers as were used for transmission) to recover the transmitted symbols. By assigning different subcarriers to different mobile devices, spectrum sharing is made possible, without requiring efficient and expensive amplifiers at the mobile units.

2.8.2 Multiple Antenna Systems

By using multiple antennas at the transmitter and/or the receiver, it is possible to significantly improve the system performance. Techniques that can be used include beamforming, receive and transmit diversity, and spatial multiplexing.

By using an array of antennas that are closely spaced with a fixed and carefully measured separation between the antennas, it is possible to exploit the deterministic phase differences between the signals on the different antennas. If an antenna array is used at the transmitter, the same signal can be sent over all the antennas, but with a slightly different delay for each one. The resulting phase differences cause the signals to combine constructively for signals propagating in some directions and destructively for other directions. As a result, it is possible to focus the transmitted signal power in any desired direction, giving a higher antenna gain in that direction. This is similar to using a directional antenna, except it is possible to quickly change the direction dynamically, just by changing the phase of the signal on each antenna element. This technique, known as *beamforming*, can also be used to simultaneously transmit two different signals to two different receivers, provided the receivers are in different directions. Beamforming can also be used at the receiver. By adjusting the phases of the signals received on the antennas, it is possible to dynamically adjust the direction in which the antenna array collects the signals. It can lock onto the signal from the desired transmitter while blocking signals arriving from transmitters in other directions.

In wireless systems where fading is a problem, it can be beneficial to place the antennas at the receiver farther apart (usually at least a few wavelengths apart). In this case the signal fading on each antenna will be nearly independent, so it is unlikely that all the received signals will be weak at the time same. The receiver could use a separate RF chain for each antenna, and assuming coherent demodulation the received baseband signal from the kth antenna would be

$$r_k(t) = \alpha_k v(t) + w_k(t) \tag{2.158}$$

where $v(t)$ is the transmitted baseband signal, and $w_k(t)$ and α_k are the noise signal and the channel attenuation due to fading, respectively, for antenna k. The noise signals will be essentially independent, and if the antennas are sufficiently far apart it is unlikely that α_k will be small for all antennas at the same time. A simple receiver could just select the signal from the "best" antenna (i.e., the antenna with the largest value of α_k) at any given time, using antenna selection. A better technique, however, is to combine the samples from all the antennas, weighting them proportionally to their attenuation in a technique known as *maximal ratio combining* (MRC). The receiver would calculate

$$r(t) = \sum_{k=1}^{N_A} \alpha_k r_k(t) = \left[\sum_{k=1}^{N_A} \alpha_k^2 \right] v(t) + \sum_{k=1}^{N_A} \alpha_k w_k(t) \tag{2.159}$$

where N_A is the number of antennas, and pass this signal to the baseband signal processor. The SNR in this case is

$$\text{SNR} = \left[\sum_{k=1}^{N_A} \alpha_k^2 \right] \frac{P_S}{P_N} \tag{2.160}$$

where P_S is the power of $v(t)$ and P_N is the noise power [the power of $w_k(t)$].

Example 2.13:   Using MRC to Increase the SNR

In a system with two receive antennas, the first receives 0.1 mVrms and the second one receives 0.13 mVrms. The noise present in the bandwidth of interest is 10 μVrms. Use maximal ratio combining and determine the overall SNR.

Solution:
The SNR from the first antenna will be 20 dB, while the second antenna will have an SNR of 22.3 dB. If antenna selection was used, then only the signal from the second antenna would be used and the SNR would be 22.3 dB. With MRC the SNR will be

$$\text{SNR} = 10 \log \frac{\alpha_1^2 P_S + \alpha_2^2 P_S}{P_N} = 10 \log \frac{(0.1 \times 10^{-3})^2 + (0.13 \times 10^{-3})^2}{(10 \times 10^{-6})^2}$$
$$= 10 \log 269 = 24.3 \text{ dB} \tag{2.161}$$

for a gain of 2 dB over antenna selection.

Use of multiple antennas at the receiver to obtain multiple copies of the transmitted signal, each with different fading, is known as *antenna diversity*, and can be very beneficial. Figure 2.48 shows the BER of a BPSK system with one, two, or three receive antennas using maximal ratio combining. Increasing the number of antennas leads to a significant improvement in performance because the BER drops more quickly as the SNR increases. The *diversity order* of a system is defined by the slope of the BER curve. With two antennas the BER drops by two orders of magnitude for every 10 dB increase in the SNR so the diversity order is 2, whereas it drops by three orders of magnitude with three antennas, giving a diversity order of 3.

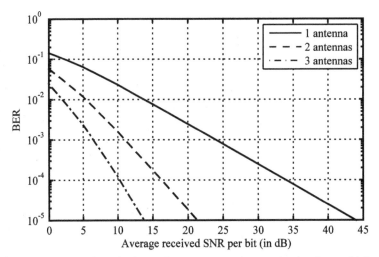

Figure 2.48 BER of BPSK with multiple receive antennas using maximal ratio combining.

It is not always possible to place multiple antennas sufficiently far apart in many mobile devices, so it is therefore not possible to get receive antenna diversity when transmitting to such devices. Instead it is possible to use multiple well-separated antennas at the transmitter, but it is more difficult to get transmit antenna diversity. If the complex channel gains are accurately known at the transmitter, the transmitter can adjust the signal phases prior to transmission so that they will combine constructively at the receiver, or the transmitter can just transmit over the "best" antenna. In the absence of channel knowledge, the transmitter can make use of a *space-time code*, such as the Alamouti code. With the Alamouti code, a transmitter with two antennas wants to transmit two symbols, v_1 and v_2, and will transmit them over two consecutive symbol periods. In the first symbol period it transmits v_1 over the first antenna and v_2 over the second antenna. In the second symbol period it transmits v_2^* over the first antenna and $-v_1^*$ over the second antenna, where * denotes the complex conjugate. By linearly combining the two received samples at the receiver, it is possible to recover both of the transmitted symbols. With the Alamouti code the transmitter sends symbols at full rate (one symbol per symbol period), and the system delivers a diversity order of 2 (which is full diversity because there are only two antennas at the transmitter). There are other space-time codes that work with more symbols, more symbol periods, and more transmit antennas, but none of these provide full rate transmission with full diversity.

If there are multiple antennas at both the transmitter and the receiver, the system is known as a *multiple-input multiple-output (MIMO) system*. With MIMO systems it is possible to make use of *spatial multiplexing*, where multiple symbols are transmitted during each symbol period, thereby significantly improving the system throughput. Some schemes, such as the vertical Bell layered space time (V-BLAST) system, do not require knowledge of the channel gains at the transmitter. Although V-BLAST is simple to implement and gives higher throughput, it does not provide diversity gains. If channel knowledge is available at the transmitter, it is possible to use precoding at the transmitter to generate multiple virtual channels between the transmitter and receiver that also provide additional diversity. Precoding techniques require elaborate signal processing involving matrix operations such as singular-value decomposition or Q-R decomposition.

2.8.3   Spread Spectrum Systems

Another way to avoid the poor performance caused by frequency flat fading is to employ a spread spectrum technique. In a spread spectrum system, a narrowband signal is intentionally spread over a much wider bandwidth than is required. By doing so, the narrowband signal, which may have been subjected to frequency flat fading, is more likely to experience frequency selective fading, which is less detrimental because only a portion of the signal is lost during deep fades. Spread spectrum systems are also more robust against narrowband interference such as intentional jamming. Better security against malicious parties eavesdropping on the communication is also provided. Because the signal power spectral density is very low (the total signal power is spread over a very wide frequency band), it is difficult for an observer to detect whether or not communication is taking place, and without knowledge of how the spreading is performed it is nearly impossible for an eavesdropper to recover the signal. The two most common spread spectrum systems in use are frequency hopping spread spectrum (FHSS) and direct sequence spread spectrum (DSSS).

With frequency hopping the narrowband signal is transmitted using a rapidly changing carrier frequency. That is, the carrier hops from frequency to frequency over time, as illustrated in Figure 2.49. During any time slot the system transmits in one frequency band, and in the next time slot the system hops to a different frequency band. The hopping sequence, which determines which frequency band to use next, is determined in a pseudo-random fashion that is known to both the transmitter and receiver. An eavesdropper that does not know the hopping pattern will not be able to track the signal because it will not know which band to tune to at any given time. There are two approaches to frequency hopping, depending on how quickly the hopping occurs. With slow frequency hopping, the carrier frequency changes more slowly than the symbol duration, while with fast frequency hopping it changes more quickly. With fast frequency hopping, each symbol is transmitted at many different frequencies, so it is unlikely that an entire symbol will be lost due to fading in some bands. However the RF system design is more complicated because the hopping occurs so rapidly. With slow hopping, it is possible that whole symbols will be lost, but because other symbols are received reliably, an error correcting coding scheme can be used to recover the lost symbols. It is worth noting that even with slow hopping it is difficult to acquire a carrier reference phase, so QAM and PSK are difficult to implement. FHSS is better used with FSK or ASK.

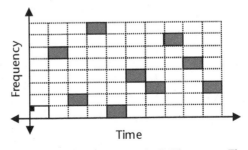

Figure 2.49   Example of frequency usage over time in FHSS systems. The shaded regions indicate which frequency band is used at any given time.

Direct sequence spread spectrum provides an alternative method for spreading the signal. It involves using a pulse shape with a very wide bandwidth. The impulse response of the filter is usually made up of a large number of very short rectangular pulses called chips. Figure 2.50 shows an example of an appropriate pulse shape for use with DSSS. In this example the pulse shape is composed of $\eta = 13$ chips, each with a duration of T_c seconds. The overall pulse duration is $T = \eta T_c$. Because the resulting signal has a bandwidth that is proportional to $1/T_c = \eta/T$ instead of $1/T$, this is a spread spectrum signal, with a bandwidth expansion factor of η. Recovery of the transmitted signal at the receiver can be performed with a matched filter, matched to the spread pulse shape.

In practice, much of the pulse shaping is performed digitally. The transmitted constellation point is repeated and multiplied by a spreading code, which is a sequence of ±1s reflecting the amplitudes of the pulse shape. For the pulse shape shown in Figure 2.50 the spreading code is [+1,−1,+1,−1,−1,+1,−1,−1,+1,+1,−1, +1,−1] so instead of just transmitting an amplitude of, say, A, for T seconds, the sequence of amplitudes [$A,−A,A,−A,−A,A,−A,−A,A,A,−A,A,−A,$] is transmitted, with each segment lasting only T_c seconds. Similarly, at the receiver a filter matched to the chip shape is used and the signal is sampled once every T_c seconds. These samples are then combined to give the output that would have been produced by a filter matched to the full pulse shape.

By spreading each symbol over a wide frequency band, the whole symbol is less likely to be lost because of fading. It is still necessary to counteract the ISI caused by frequency-selective fading, but a simple RAKE receiver, which involves using many matched filters or discrete time correlators with different delays (in multiples of the chip duration), can be used. The outputs of these filters are combined to recover all the received energy for each transmitted symbol. DSSS systems also provide good protection against narrowband interference. When the received signal is despread by the matched filter, this has the effect of spreading the narrowband interference, so it is less harmful.

DSSS systems are also useful for sharing spectrum over multiple users. In a *code division multiple access* (CDMA) scheme, each user is assigned a different spreading code, with each spreading code orthogonal (or nearly orthogonal) to each other. As a result, a receiver that is matched to one spreading code will block signals generated with other spreading codes, allowing only the desired signal to be received. In the downlink of a multiuser CDMA system, perfectly orthogonal spreading codes can be used because there is only one transmitter (the base station) so it is easy to

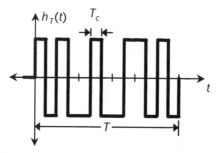

Figure 2.50 Example DSSS pulse shape, where T is the pulse duration and T_c is the chip duration.

keep the transmissions to all the remote users perfectly synchronized. However, such synchronization is difficult in the uplink, because all the transmitting users would need to adjust their transmit times to ensure that their signals arrive synchronously at the base station. Because perfectly orthogonal spreading codes are only orthogonal when received synchronously, and are highly correlated otherwise, nearly orthogonal spreading codes are used in the uplink. These spreading codes are attractive because although they are slightly correlated, the correlation does not change when they are not synchronized. This means that users are free to transmit whenever they want, without having to start transmitting at carefully prescribed times. However, because the codes are not perfectly orthogonal, if the signal from one user is received at a much higher level than the one from a different user, the nonorthogonal nature of the spreading codes can lead to high levels of multi-user interference. This is known as the *near-far problem*. It can be reduced by using a transmit power control loop to ensure that the received signal power for all users is roughly equal (that is, nearby users transmit at a lower power than distant ones).

2.8.4   Error Control Coding

Error control coding is really useful for reducing the SNR. An additional important technique for improving the reliability of the communication system is to employ error control coding. An encoder at the transmitter can add carefully controlled redundancy to the data prior to transmission, and a corresponding decoder at the receiver can exploit this redundancy to detect and/or correct transmission errors.

In general, an error control coding scheme involves taking a *message word*, consisting of k message bits, and mapping it to a *code word*, consisting of $n > k$ code bits, prior to transmission. The extra $m = n - k$ bits appended to the message word to give the code word are referred to as the *parity bits*, and these bits depend only on the message word. Instead of transmitting the original message word, the transmitter sends the code word, using a regular modulation scheme.

Perhaps the most simple error control coding scheme is the use of a single parity bit. With an even parity bit code, an extra bit (the parity bit) is added at the end the message word (where typically $k \leq 8$) such that the number of bits with a value of one in the resulting codeword of $n = k + 1$ bits will be even. For example, the sequence 1011 will be encoded as 10111, where the final bit is the parity bit. The receiver, after receiving all n bits, can count the number of ones, and if this number is odd then it knows that an error as occurred. With this coding scheme, the receiver can detect the occurrence of any single bit error out of the n transmitted bits. It can also detect whenever an odd number of bit errors occurs, but if the probability of a bit error is low then it is much less likely that multiple bits will be in error if n is small. However, if an even number of bit errors occurs, then the receiver will not be able to detect that errors have occurred. Single parity bit codes are widely used in short-range wired communication systems such as SCSI and PCI buses, but not so much in wireless systems.

More common in wireless systems are cyclic redundancy check (CRC) codes. These codes work with much longer messages, and add multiple parity bits. Typically m is 8 to 16 bits, but may be as long as 32 bits, depending on the code, while k can be several thousand bits, if needed. CRC codes have the important property that a very simple encoding circuit at the transmitter can be used to calculate the parity bits (the CRC bits) based on the message to transmit, and another simple circuit at the receiver can be used to verify if the CRC bits are valid for the received message. If the CRC does not match what is expected based on the received message, then the receiver is able to detect that transmission errors have occurred. It is worth noting that the single parity bit code is a special case of a CRC code with $m = 1$. Provided that m is sufficiently large and k is not too big, it is extremely unlikely that any errors that occur during transmission will not be detected.

If errors are detected, the receiver can ask for a retransmission of the message, and it can continue doing so until the message is received correctly (or at least until no errors are detected). This type of retransmission scheme, known as *automatic repeat request* (ARQ), is widely used in wireless systems (and in the Internet). Although it can give a much lower probability of error, it does lead to a reduction in throughput, because there is overhead involved in transmitting the parity bits (although this is negligible if $k \gg m$), and overhead involved in the retransmissions when errors are detected.

In some wireless communication systems it is not possible to perform retransmissions. For example, in a broadcast environment, where multiple receivers are trying to receive the same signal simultaneously, the number of needed retransmissions would become excessively large because there would be a high probability that at least one receiver would require a retransmission for every transmitted message. Also, in situations where the propagation delay is very large, the transmitter would need a lot of memory to store previously transmitted messages while waiting to learn if a retransmission is required. For example, the propagation delay from Jupiter to the Earth is about 45 minutes, so a deep space probe would have to store 90 minutes worth of transmitted data in case a retransmission is requested. In such cases it is necessary to use an *error-correcting code*.

With an error-correcting code, the transmitter adds a much higher fraction of redundancy to the transmitted message, allowing for error correction as well as detection at the receiver. This is useful not only for cases where retransmissions are impossible, but also for cases where it is desirable to reduce the probability that retransmissions are required. Most modern communication systems use error correcting codes.

An illustrative example of an error correcting code is the (6,3) shortened Hamming code, where message words of $k = 3$ bits are mapped to code words of $n = 6$ bits at the transmitter, according to the mapping rule shown in Table 2.3. For example, the message 100 is mapped to the code word 100011, which is transmitted. If the received code word is not in the table, then the receiver knows that transmission errors have occurred. It can either request a retransmission, or it can attempt to correct the errors. Because it is not likely that many errors occurred, the optimal decoding strategy is to find the code word in the table that is "closest" to the received word, where closeness is measured in terms of the *Hamming distance*. The Hamming distance between two sequences of bits is the number of bit positions in which the two sequences differ. So, for example, if 111110 was received, the receiver would assume that 110110 was transmitted, since 111110 and 110110 differ in only one bit position (i.e., the Hamming distance is 1), whereas the Hamming distance between 111110 and the other code words in the table is 2 or more. As 110110 corresponds to the message 110, the received message would be 110.

Table 2.3 Code Words for the (6,3) Shortened Hamming Code

Message	Code Word
000	000000
001	001110
010	010101
011	011011
100	100011
101	101101
110	110110
111	111000

Although this simple Hamming code is useful for illustration the operation of error correcting codes, there are much better codes available. Some examples include convolutional codes, turbo codes, and low density parity check (LDPC) codes. These codes generally operate on much longer messages (from several hundred to thousands of bits), yet have simple encoders and decoders.

With any error correcting code, it is still possible that errors will remain after error correction has been attempted. That is, the code word that is most likely to have been transmitted is not necessarily the code word that was transmitted. One important factor affecting the performance of a code is the *code rate*, which is defined as the fraction of message bits per code bit (i.e., $R = k/n$), and in general, the smaller the rate, the lower the BER. However, this comes with a corresponding reduction in the throughput, by a factor of R, so in most applications only codes with rates greater than about 1/3 are used.

It is not possible to find exact expressions for the BER for most practical error-correcting codes. Instead the BER is usually found through computer simulation, and plotted as a function of the energy per message bit over the noise power spectral density. In Figure 2.51, the BER of the convolutional codes used in communication systems based on the IEEE 802.11 standard are shown, when the code bits are transmitted using BPSK. Also shown is the performance of uncoded BPSK. With

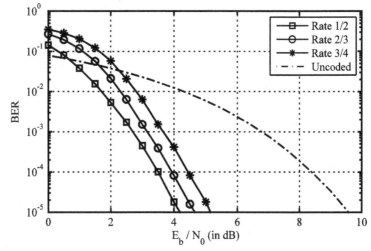

Figure 2.51 Probability of bit error for convolutionally encoded BPSK.

the rate 1/2 convolutional code, we need E_b/N_0 of only 4.25 dB to achieve a BER of 10^{-5}, which is much less than the 9.5 dB needed for uncoded BPSK. However, because of the overhead in transmitting the parity bits, we have a 50% reduction in the throughput as well. We see that the higher-rate codes require slightly higher values of E_b/N_0, but also suffer from less of a reduction in throughput.

As a final note, it is important to remember that E_b in Figure 2.51 is the energy per message bit, not the energy per code bit, and it is necessary to take this into account when determining the SNR required to achieve a given BER. The energy per code bit is related to the energy per message bit by $E_{cb} = RE_b$, and if $\log_2 M$ code bits are transmitted per symbol, depending on the modulation scheme, then the SNR is

$$\gamma_s = \frac{E_s}{N_0} = \frac{E_{cb}\log_2 M}{N_0} = \frac{E_b}{N_0} R \log_2 M \qquad (2.162)$$

which is similar to (2.70) except for the inclusion of the factor R.

Example 2.14:   Required SNR for Coded BPSK

Find the SNR required to achieve a BER of 10^{-4} when BPSK is used with the code giving the performance shown in Figure 2.51. Consider the rate 1/2 code, the rate 3/4 code, and the uncoded case.

Solution:
For the rate 1/2 code to achieve a BER of 10^{-4}, from Figure 2.51 we need E_b/N_0 = 3.75 dB. Because $R = 1/2$ and $M = 2$, from (2.162) the required SNR is

$$\gamma_s = 10^{3.75/10}\frac{1}{2}\log_2 2 = 1.19 = 0.74 \text{ dB} \qquad (2.163)$$

For the rate 3/4 code we need E_b/N_0 = 4.2 dB and γ_s = 2.95 dB, and for the uncoded case we need E_b/N_0 = 8.5 dB and γ_s = 8.5 dB. We therefore get a gain of 7.75 dB by using the rate 1/2 code instead of the uncoded system, at the expense of a 50% reduction in throughput.

2.9   Summary

This chapter introduced the most important principles of digital communication theory, including signal space representation of signals and the design of optimum receivers with matched filters. The spectral characteristics of several common modulation schemes were explored, and the performance of these schemes in terms of the probability of error was analyzed. The fundamental signal impairments, including fading, arising from transmission over the wireless medium were discussed. Several more advanced techniques for improving the reliability of communication over wireless channels, including OFDM, MIMO systems, spread spectrum systems, and error control coding, were introduced.

There are many excellent textbooks that cover these topics in greater detail. Some particularly useful ones are listed here. More detailed descriptions of the theoretical aspects of digital communication, suitable for a graduate level, can be found in [1, 2]. These books also discuss error analysis, which is also comprehensively covered in [3]. More introductory-level material on communication systems is given in [4–6]. A good description of the signal impairments in wireless communication channels can be found in [7], while details on various signal processing techniques for wireless communication systems are provided in [8, 9]. The theoretical aspects of error control coding are given in [10].

References

[1]   Proakis, J. G., *Digital Communications*, 4th ed., New York: McGraw-Hill, 2001.

[2]   Wozencraft, J. M, and I. M. Jacobs, *Principles of Communication Engineering*, New York: John Wiley & Sons, 1965.

[3]   Simon, M. K., S. M. Hinedi, and W. C. Lindsey, *Digital Communication Techniques*, Upper Saddle River, NJ: Prentice Hall, 1995.

[4]   Sklar, B., *Digital Communications*, 2nd ed., Upper Saddle River, NJ: Prentice Hall, 2001.

[5]   Haykin, S., *Communication Systems*, 4th ed., New York: John Wiley & Sons, 2001.

[6]   Couch, L.W., *Digital and Analog Communication Systems*, Upper Saddle River, NJ: Pearson Prentice Hall, 2007.

[7]   Rappaport, T. S., *Wireless Communication Systems*, Upper Saddle River, NJ: Prentice-Hall, 1996.

[8]   Tse, D., and P. Viswanath, *Fundamentals of Wireless Communication*, Cambridge, U.K.: Cambridge University Press, 2005.

[9]   Wang, X., and H. V. Poor, *Wireless Communication Systems*, Upper Saddle River, NJ: Pearson Prentice Hall, 2004.

[10] Lin, S., and D. J. Costello, Jr., *Error Control Coding*, 2nd ed., Upper Saddle River, NJ: Pearson Prentice Hall, 2004.

CHAPTER 3

Basic RF Design Concepts and Building Blocks

3.1 Introduction

In this chapter some general issues in RF design will be considered. Nonidealities including noise and linearity will be discussed. An ideal circuit, for example, an ideal amplifier, produces a perfect copy of the input signal at the output. In a real circuit the amplifier will add both noise and distortion to that waveform. Noise, which is present in all resistors and active devices, limits the minimum detectable signal in a radio. At the other amplitude extreme, nonlinearities in the circuit blocks will cause the output signal to become distorted, limiting the maximum signal amplitude.

At the system level, specifications for linearity, and noise as well as many other parameters, must be determined before the circuit can be designed. In this chapter, some of these system issues will be looked at in more detail. To specify radio-frequency integrated circuits with realistic performance, the impact of noise on minimum detectable signals and the effect of nonlinearity on distortion need to be understood. Knowledge of noise floors and distortion will be used to understand the requirements for circuit parameters. Finally, the chapter will conclude by studying some common RF blocks used in communications.

3.2 Gain

Perhaps one of the most fundamental properties of many basic building blocks considered in this book will be gain. Often blocks are designed to sense some property of the input signal (the input voltage, current, or power) and amplify this or some other property at the output. There are many different types of gain that can be defined for a block, for instance there is power gain:

$$G = \frac{P_{\text{out}}}{P_{\text{in}}} \tag{3.1}$$

which is defined as the power output from a block divided by the power delivered to the input of the block. Some common gain definitions are listed in Table 3.1.

Table 3.1 Summary of Some Different Gain Definitions

Type of Gain	Equation	Description
Power gain	$G = 10\log\left(\dfrac{P_{out}}{P_{in}}\right)$	The ratio of ouptut power to input power.
Voltage gain	$A_v = 20\log\left(\dfrac{V_{out}}{V_{in}}\right)$	The ratio of the output voltage to the input voltage.
Current gain	$A_i = 20\log\left(\dfrac{i_{out}}{i_{in}}\right)$	The ratio of the output current to the input current.
Maximum available power gain	$G_A = 10\log\left(\dfrac{P_{out\_av}}{P_{in\_av}}\right)$	The ratio of maximum available output power to the maximum available input power.
Transducer power gain	$G_T = 10\log\left(\dfrac{P_{out}}{P_{in\_av}}\right)$	The ratio of the output power to the maximum available input power.
Open ciruit voltage gain	$A_{v\_oc} = 20\log\left(\dfrac{v_{out\_OC}}{v_{in}}\right)$	The ratio of the output voltage to the input voltage when the circuit is loaded with only an open circuit load.

3.3 Noise

Signal detection is more difficult in the presence of noise. In addition to the desired signal, the receiver is also picking up noise from the rest of the universe. Any matter above zero Kelvin contains thermal energy. This thermal energy moves atoms and electrons around in a random way, leading to random currents in circuits, which are also seen as noise. Noise can also come from man-made sources such as microwave ovens, cell phones, pagers, and radio antennas. RF designers are mostly concerned with how much noise is being added by the circuits in the transceiver. At the input to the receiver, there will be some noise power present that defines the noise floor. The minimum detectable signal must be higher than the noise floor by some signal-to-noise ratio (SNR) to detect signals reliably and to compensate for additional noise added by circuitry. These concepts will be described in the following sections.

To find the total noise due to a number of sources, the relationship of the sources with each other has to be considered. The most common assumption is that all noise sources are random and have no relationship with each other, so they are said to be uncorrelated. In such a case, noise power is added instead of noise voltage. Similarly, if noise at different frequencies is uncorrelated, noise power is added. We note that signals, like noise, can also be uncorrelated, for example, signals at different unrelated frequencies. In such a case, one finds the total output signal by adding the powers. In contrast, if two sources are correlated, the voltages can be added. As an example, correlated noise is seen at the outputs of two separate paths that have the same origin.

3.3.1 Thermal Noise

One of the most common noise sources in a circuit is a resistor. Noise in resistors is generated by thermal energy causing random electron motion [1, 2]. The thermal noise spectral density in a resistor is given by

$$N_{resistor} = 4kTR \tag{3.2}$$

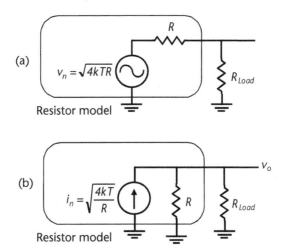

Figure 3.1 Resistor noise model: (a) with a voltage source and (b) with a current source.

where T is the temperature in Kelvin of the resistor, k is Boltzmann's constant (1.38×10^{-23} Joules/K), and R is the value of the resistor. Noise power spectral density has the units of V^2/Hz (power spectral density). To determine how much power a resistor produces in a finite bandwidth, simply multiply (3.2) by the bandwidth of interest Δf:

$$v_n^2 = 4kTR\Delta f \qquad (3.3)$$

where v_n is the rms value of the noise voltage in the bandwidth Δf. This can also be written equivalently as a noise current rather than a noise voltage:

$$i_n^2 = \frac{4kT\Delta f}{R} \qquad (3.4)$$

Thermal noise is white noise, meaning it has a constant power spectral density with respect to frequency (valid up to approximately 6,000 GHz) [3]. The model for noise in a resistor is shown in Figure 3.1.

3.3.2 Available Noise Power

Maximum noise power of a resistor k is transferred to the load when R_{LOAD} is equal to R. Then v_o is equal to $v_n/2$. The output power spectral density P_0 is then given by

$$P_0 = \frac{v_0^2}{R} = \frac{v_n^2}{4R} = kT \qquad (3.5)$$

Thus, the available power is kT, independent of resistor size. Note that kT is in watts per hertz, which is a power density. To get total power out, P_{out} in watts, multiply P_0 by the bandwidth with the result that

$$P_{\text{out}} = kTB \qquad (3.6)$$

3.3.3 Available Noise Power from an Antenna

The noise power from an antenna with input resistance R is equivalent to that of a physical resistor with the same value [4]. Thus, as in the previous section, the available noise power from an antenna is given by

$$P_{\text{available}} = kT = 4 \cdot 10^{-21} \, \text{W/Hz} \qquad (3.7)$$

at $T = 290\text{K}$, or in dBm/Hz:

$$P_{\text{available}} = 10 \log_{10} \left(\frac{4 \times 10^{-21}}{1 \times 10^{-3}} \right) = -174 \, \text{dBm/Hz} \qquad (3.8)$$

Using 290K as the temperature of the resistor when modeling the antenna is appropriate for cell phone applications where the antenna is pointed at the horizon. However, if the antenna were pointed at the sky, the equivalent noise temperature would be much lower, more typically 50K [5].

For any receiver required to receive a given bandwidth of signal, the minimum detectable signal can now be determined. As can be seen from (3.6), the noise floor depends on the bandwidth. For example, with a bandwidth of 200 kHz, the noise floor is

$$\text{Noise Floor} = kTB = 4 \times 10^{-21} \times 200,000 = 8 \times 10^{-16} \, \text{W} \qquad (3.9)$$

More commonly, the noise floor would be expressed in dBm, as in the following for the same example as before:

$$\text{Noise Floor} = -174 \, \text{dBm/Hz} + 10 \log_{10}(200,000) = -121 \, \text{dBm} \qquad (3.10)$$

Thus, signal-to-noise ratio (SNR) can now also be formally defined. If the signal has a power of S, then the SNR is

$$\text{SNR} = \frac{S}{\text{Noise Floor}} \qquad (3.11)$$

Thus, if the electronics added no noise and if the detector required an SNR of 0 dB, then a signal at –121 dBm could just be detected. The minimum detectable signal in a receiver is also referred to as the receiver sensitivity. However, the SNR required to detect bits reliably (for example, at a BER of 10^{-3}) is typically not 0 dB. The actual required SNR depends on a variety of factors, such as modulation scheme, bit rate, energy per bit, IF filter bandwidth, detection method (for example, synchronous or not), and interference levels. Such calculations [5, 6] will be discussed in other chapters. Typical results for a bit error rate of 10^{-3} are about 7 dB for QPSK, about 12 dB for 16 QAM, and about 17 dB for 64 QAM, although often higher numbers are quoted to leave a safety margin. It should be noted that a lower BER is often required (for example, 10^{-6}), resulting in a SNR requirement of 11 dB or more for QPSK. Thus, the input signal level must be above the noise floor level by at least this amount. Consequently, the minimum detectable signal level in a 200-kHz bandwidth is more like –114 to –110 dBm (assuming that no noise is added by the electronics).

3.3.4 The Concept of Noise Figure

Noise added by electronics will directly add to the noise from the input. Thus, for reliable detection, the previously calculated minimum detectable signal level must be modified to include the noise from the active circuitry. Noise from the electronics is described by noise factor F, which is a measure of how much the SNR is degraded through the system. By noting that:

$$S_o = G \cdot S_i \qquad (3.12)$$

where S_i is the input signal power, S_o is the output signal power, and G is the available power gain S_o/S_i, the following equation for noise factor can be derived:

$$F = \frac{\text{SNR}_i}{\text{SNR}_o} = \frac{S_i/N_{i(\text{source})}}{S_o/N_{o(\text{total})}} = \frac{S_i/N_{i(\text{source})}}{S_i \cdot G/N_{o(\text{total})}} = \frac{N_{o(\text{total})}}{G \cdot N_{i(\text{source})}} \qquad (3.13)$$

where $N_{o(\text{total})}$ is the total noise power at the output and $N_{i(\text{source})}$ is the noise at the input due to the source. If $N_{o(\text{source})}$ is the noise power at the output originating at the source, and $N_{o(\text{added})}$ is the noise power at the output added by the electronic circuitry. Then

$$N_{o(\text{total})} = N_{o(\text{source})} + N_{o(\text{added})} \qquad (3.14)$$

the noise factor can be written in several useful alternative forms:

$$F = \frac{N_{o(\text{total})}}{G \cdot N_{i(\text{source})}} = \frac{N_{o(\text{total})}}{N_{o(\text{source})}} = \frac{N_{o(\text{source})} + N_{o(\text{added})}}{N_{o(\text{source})}} = 1 + \frac{N_{o(\text{added})}}{N_{o(\text{source})}} \qquad (3.15)$$

This shows that the minimum possible noise factor, which occurs if the electronics adds no noise, is equal to 1. The noise figure, NF, is related to noise factor, F, by

$$\text{NF} = 10 \log_{10} F \qquad (3.16)$$

Thus, while the noise factor is at least 1, the noise figure is at least 0 dB. In other words, an electronic system that adds no noise has a noise figure of 0 dB.

In the receiver chain, for components with loss (such as switches and filters), the noise figure is equal to the attenuation of the signal. For example, a filter with 3 dB of loss has a noise figure of 3 dB. This is explained by noting that output noise is approximately equal to input noise, but the signal is attenuated by 3 dB. Thus, there has been a degradation of SNR by 3 dB.

Example 3.1: Noise Calculations

Figure 3.2 shows a 50Ω source resistance loaded with 50Ω. Determine how much noise voltage per unit bandwidth is present at the output. Also find the noise factor, assuming that R_L does not contribute to the noise factor, and compare to the case where R_L does contribute to the noise factor.

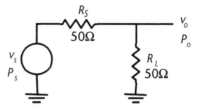

Figure 3.2 Simple circuit used for noise calculations.

Solution:
The rms noise voltage from the 50Ω source is $\sqrt{4kTR} = 0.894\,\text{nV}/\sqrt{\text{Hz}}$ at a temperature of 290K, which, after the voltage divider, becomes one-half of this value or $v_o = 0.447\,\text{nV}/\sqrt{\text{Hz}}$.

The complete available power from the source is delivered to the load. In this case,

$$P_o = P_{in(available)} = kT = 4 \cdot 10^{-21}$$

At the output, the complete noise power (available) appears and so, if R_L is noiseless, the noise factor is 1. However, if R_L has noise of $\sqrt{4kTR_L}\,\text{V}/\sqrt{\text{Hz}}$, then at the output, the total noise power is $2kT$ where kT is from R_S and kT is from R_L. Therefore, for a resistively matched circuit, the noise figure is 3 dB. Note that the output noise voltage is $0.45\,\text{nV}/\sqrt{\text{Hz}}$ from each resistor for a total of $\sqrt{2} \times 0.45\,\text{nV}/\sqrt{\text{Hz}} = 0.636\,\text{nV}/\sqrt{\text{Hz}}$ (with noise, the power adds because the noise voltage is uncorrelated).

Example 3.2: Noise Calculation with Gain Stages

In this example, in Figure 3.3, a voltage gain of 20 has been added to the original circuit of Figure 3.2. All resistor values are still 50Ω. Determine the noise at the output of the circuit due to all resistors and then determine the circuit noise figure and signal-to-noise ratio assuming a 1-MHz bandwidth and the input is a 1-V sine wave.

Solution:
In this example, at v_x the noise is still due to only R_S and R_2. As in the previous example, the noise at this point is $0.636\,\text{nV}/\sqrt{\text{Hz}}$. The signal at this point is 0.5V, thus at point v_y, the signal is 10V and the noise due to the two input resistors R_S and R_2 is $0.636 \times 20 = 12.72\,\text{nV}/\sqrt{\text{Hz}}$. At the output, the signal and noise from the input sources, as well as the noise from the two output resistors, all see a voltage divider. Thus, one can calculate the individual components. For the combination of R_S and R_2, one obtains

$$v_{R_S + R_2} = 0.5 \times 12.72\,\text{nV}/\sqrt{\text{Hz}} = 6.36\,\text{nV}/\sqrt{\text{Hz}}$$

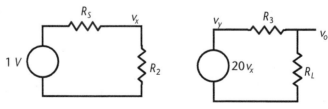

Figure 3.3 Noise calculation with a gain stage.

The noise from the source can be determined from this equation:

$$v_{R_S} = \frac{6.36 \text{ nV}/\sqrt{\text{Hz}}}{\sqrt{2}} = 4.5 \text{ nV}/\sqrt{\text{Hz}}$$

For the other resistors, the voltage is

$$v_{R_3} = 0.5 \times 0.9 \text{nV}/\sqrt{\text{Hz}} = 0.45 \text{nV}/\sqrt{\text{Hz}}$$

$$v_{R_L} = 0.5 \times 0.9 \text{nV}/\sqrt{\text{Hz}} = 0.45 \text{nV}/\sqrt{\text{Hz}}$$

Total output noise is given by

$$v_{no(total)} = \sqrt{v_{R_S+R_2}^2 + v_{R_3}^2 + v_{R_L}^2}$$

$$= \sqrt{6.36^2 + 0.45^2 + 0.45^2} \left(\text{nV}/\sqrt{\text{Hz}}\right) = 6.392 \text{ nV}/\sqrt{\text{Hz}}$$

Therefore, the noise figure can now be determined:

$$NF = 10\log F = 10\log\left(\frac{N_{o(total)}}{N_{o(source)}}\right) = 10\log\left(\frac{6.392}{4.5}\right)^2 = 10\log(1.417)^2 = 3.03 \text{ dB}$$

Since the output voltage also sees a voltage divider of 1/2, it has a value of 5V. Thus, the SNR is

$$SNR = 20\log\left(\frac{5}{\frac{6.392 \text{ nV}}{\sqrt{\text{Hz}}} \cdot \sqrt{1 \text{ MHz}}}\right) = 117.9 \text{ dB}$$

This example illustrates that noise from the source and amplifier input resistance are the dominant noise sources in the circuit. Each resistor at the input provided 4.5 nV/$\sqrt{\text{H}}$z, while the two resistors behind the amplifier each only contribute 0.45 nV/$\sqrt{\text{H}}$z. Thus, as explained earlier, after a gain stage, noise is less important.

Example 3.3:   Effect of Impedance Mismatch on Noise Figure

Find the noise figure of Example 3.2 again, but now assume that $R_2 = 500\Omega$.

Solution:
As before, the output noise due to the resistors is as follows:

$$v_{no(R_s)} = \left(0.9 \times \frac{500}{550} \times 20 \times 0.5\right) \text{nV}/\sqrt{\text{Hz}} = 8.181 \text{ nV}/\sqrt{\text{Hz}}$$

where 500/550 accounts for the voltage division from the noise source to the node v_x.

$$v_{no(R_2)} = \left(0.9 \times \sqrt{10} \times \frac{50}{550} \times 20 \times 0.5\right) \text{nV}/\sqrt{\text{Hz}} = 2.587 \text{ nV}/\sqrt{\text{Hz}}$$

where the $\sqrt{10}$ accounts for the higher noise in a 500Ω resistor compared to a 50Ω resistor.

$$v_{no(R_3)} = \left(0.9 \times 0.5\right) \text{nV}/\sqrt{\text{Hz}} = 0.45 \text{ nV}/\sqrt{\text{Hz}}$$

$$v_{no(R_L)} = \left(0.9 \times 0.5\right) \text{nV}/\sqrt{\text{Hz}} = 0.45 \text{ nV}/\sqrt{\text{Hz}}$$

The total output noise voltage is

$$v_{no(total)} = \sqrt{v_{R_S}^2 + v_{R_2}^2 + v_{R_3}^2 + v_{R_L}^2} = \left(\sqrt{8.181^2 + 2.587^2 + 0.45^2 + 0.45^2}\right) nV/\sqrt{Hz}$$

$$= 8.604 \ nV/\sqrt{Hz}$$

The noise figure is

$$NF = 10\log\left(\frac{N_{o(total)}}{N_{o(source)}}\right) = 10\log\left(\frac{8.604}{8.181}\right)^2 = 0.438 \ dB$$

Note that this circuit is unmatched at the input. This example illustrates that a mismatched circuit may have better noise performance than a matched one. However, this assumes that it is possible to build a voltage amplifier that requires little power at the input. This may be possible on an IC. However, if transmission lines are included, power transfer will suffer. A matching circuit may need to be added.

3.3.5 Phase Noise

Radios use reference tones to perform frequency conversion. Ideally, these tones would be perfect and have energy at only the desired frequency. Unfortunately, any real signal source will emit energy at other frequencies as well. Local oscillator noise performance is usually classified in terms of phase noise, which is a measure of how much the output diverges from an ideal impulse function in the frequency domain. We are primarily concerned with noise that causes fluctuations in the phase of the output rather than noise that causes amplitude fluctuations in the tone, since the output typically has a fixed, limited amplitude. The output signal of a reference tone can be described as

$$v_{out}(t) = V_0 \cos(\omega_{LO}t + \varphi_n(t)) \tag{3.17}$$

Here, $\omega_{LO}t$ is the desired phase of the output and $\varphi_n(t)$ are random fluctuations in the phase of the output due to any one of a number of sources. Phase noise is often quoted in units of dBc/Hz while timing jitter is often quoted in units of rad²/Hz. The phase fluctuation term $\varphi_n(t)$ may be random phase noise or discrete spurious tones, as shown in Figure 3.4. As can be seen in this figure phase noise is measured by comparing the power at the desired frequency to the noise power present in a 1-Hz bandwidth at some frequency offset from the carrier.

Assume that the phase fluctuation is of a sinusoidal form as

$$\varphi_n(t) = \varphi_p \sin(\omega_m t) \tag{3.18}$$

where φ_p is the peak phase fluctuation and ω_m is the offset frequency from the carrier. Substituting (3.18) into (3.17) gives

$$v_{out}(t) = V_0 \cos\left[\omega_{LO}t + \varphi_p \sin(\omega_m t)\right]$$

$$= V_0\left[\cos(\omega_{LO}t)\cos(\varphi_p \sin(\omega_m t)) - \sin(\omega_{LO}t)\sin(\varphi_p \sin(\omega_m t))\right] \tag{3.19}$$

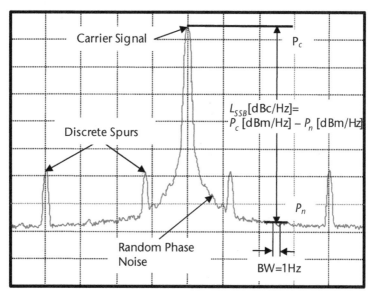

Figure 3.4   An example of phase noise and spurs observed using a spectrum analyzer.

For a small phase fluctuation, φ_p, the above equation can be simplified as:

$$v_0\left(t\right) = V_0\left[\cos\left(\omega_{LO}t\right) - \varphi_p\sin\left(\omega_m t\right)\sin\left(\omega_{LO}t\right)\right]$$

$$= V_0\left[\cos\left(\omega_{LO}t\right) - \frac{\varphi_p}{2}\left[\cos\left(\left[\omega_{LO} - \omega_m\right]t\right) - \cos\left(\left[\omega_{LO} + \omega_m\right]t\right)\right]\right] \qquad (3.20)$$

It is now evident that the phase-modulated signal includes the carrier signal tone and two symmetric sidebands at some offset frequency, as shown in Figure 3.4. A spectrum analyzer measures the phase-noise power in dBm/Hz, but often phase noise is reported relative to the carrier power as:

$$\varphi_n^2\left(\Delta\omega\right) = \frac{Noise\left(\omega_{LO} + \Delta\omega\right)}{P_{carrier}\left(\omega_{LO}\right)} \qquad (3.21)$$

where *Noise* is the noise power in a 1-Hz bandwidth and $P_{carrier}$ is the power of the carrier or LO tone at the frequency at which the synthesizer is operating. In this form, phase noise has the units of rad²/Hz. Often this is quoted as so many decibels down from the carrier or in dBc/Hz. To further complicate this, both single-sideband and double-sideband phase noise can be defined. Single-sideband (SSB) phase noise is defined as the ratio of power in one phase modulation sideband per hertz of bandwidth, at an offset $\Delta\omega$ away from the carrier, to the total signal power. The SSB phase noise power spectral density (PSD) to carrier ratio, in units of [dBc/Hz], is defined as

$$PN_{SSB}\left(\Delta\omega\right) = 10\log\left[\frac{Noise\left(\omega_{LO} + \Delta\omega\right)}{P_{carrier}\left(\omega_{LO}\right)}\right] \qquad (3.22)$$

Combining (3.20) into (3.22) this equation can be rewritten as

$$PN_{SSB}(\Delta\omega) = 10\log\left[\frac{\frac{1}{2}\left(\frac{V_0\varphi_p}{2}\right)^2}{\frac{1}{2}V_0^2}\right] = 10\log\left[\frac{\varphi_p^2}{2}\right] = 10\log\left[\frac{\varphi_{rms}^2}{2}\right] \qquad (3.23)$$

where φ_{rms}^2 is the rms phase-noise power density in units of rad²/Hz. Note that single-sideband phase noise is by far the most common type reported and often it is not specified as SSB, but rather simply reported as phase noise. However, alternatively, double sideband phase noise is given by

$$PN_{DSB}(\Delta\omega) = 10\log\left[\frac{Noise(\omega_{LO}+\Delta\omega) + Noise(\omega_{LO}-\Delta\omega)}{P_{carrier}(\omega_{LO})}\right] = 10\log\left[\varphi_{rms}^2\right] \quad (3.24)$$

From either the single-sideband or double-sideband phase noise, the rms jitter can be obtained as

$$\varphi_{rms}(\Delta f) = \frac{180}{\pi}\sqrt{10^{\frac{PN_{DSB(\Delta f)}}{10}}} = \frac{180\sqrt{2}}{\pi}\sqrt{10^{\frac{PN_{SSB(\Delta f)}}{10}}} \left[\deg/\sqrt{Hz}\right] \qquad (3.25)$$

It is also quite common to quote integrated phase noise. The rms integrated phase noise of a synthesizer is given by

$$IntPN_{rms} = \sqrt{\int_{\Delta f_1}^{\Delta f_2} \varphi_{rms}^2(f)\,df} \qquad (3.26)$$

The limits of integration are usually the offsets corresponding to the lower and upper frequencies of the bandwidth of the information being transmitted.

In addition, it should be noted that dividing or multiplying a signal in the frequency domain also multiplies or divides the phase noise. Thus, if a signal frequency is multiplied or divided by N then the output tone will be given by:

$$v_{out}(t) = V_0\cos\left(N\omega_{LO}t + N\varphi_n(t)\right)$$

$$v_{out}(t) = V_0\cos\left(\frac{\omega_{LO}t}{N} + \frac{\varphi_n(t)}{N}\right) \qquad (3.27)$$

and thus, the phase noise is related by

$$\varphi_{rms}^2\left(N\omega_{LO}+\Delta\omega\right) = N^2\cdot\varphi_{rms}^2\left(\omega_{LO}+\Delta\omega\right)$$

$$\varphi_{rms}^2\left(\frac{\omega_{LO}}{N}+\Delta\omega\right) = \frac{\varphi_{rms}^2\left(\omega_{LO}+\Delta\omega\right)}{N^2} \qquad (3.28)$$

Note this assumes that the circuit that did the frequency translation is noiseless. Also, note that the phase noise is scaled by N^2 rather than N to get units of V^2 rather than noise voltage.

3.4   Linearity and Distortion in RF Circuits

In an ideal system, the output is linearly related to the input. However, in any real device the transfer function is usually a lot more complicated. This can be due to active or passive devices in the circuit, or the signal swing being limited by the power supply rails. Unavoidably, the gain curve for any component is never a perfectly straight line, as illustrated in Figure 3.5.

The resulting waveforms can appear as shown in Figure 3.6. For amplifier saturation, typically the top and bottom portions of the waveform are clipped equally, as shown in Figure 3.6(b). However, if the circuit is not biased at the midpoint of the two clipping levels, then clipping can be nonsymmetrical as shown in Figure 3.6(c). Here clipping occurs first on the positive voltage peaks before any distortion occurs at the lower voltage extreme of the waveform.

3.4.1   Power Series Expansion

Mathematically, any nonlinear transfer function can be written as a series expansion of power terms unless the system contains memory, in which case a Volterra series is required [8, 9]:

$$v_{out} = k_0 + k_1 v_{in} + k_2 v_{in}^2 + k_3 v_{in}^3 + \ldots \qquad (3.29)$$

To describe the nonlinearity perfectly, an infinite number of terms are required; however, in many practical circuits, the first three terms are sufficient to characterize the circuit with a fair degree of accuracy.

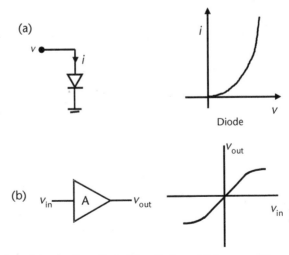

Figure 3.5   Illustration of the nonlinearity in (a) a diode and (b) an amplifier.

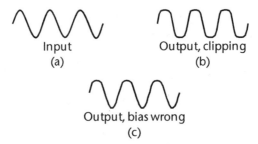

Figure 3.6 Distorted output waveforms: (a) input; (b) output, clipping; and (c) output, bias wrong.

Symmetrical saturation as shown in Figure 3.5(b) can be modeled with odd-order terms, for example:

$$y = x - \frac{1}{10} x^3 \tag{3.30}$$

looks like Figure 3.7. As another example, an exponential nonlinearity as shown in Figure 3.5(a) has the form:

$$x + \frac{x^2}{2!} + \frac{x^3}{3!} + \dots \tag{3.31}$$

which contains both even and odd power terms, because it does not have symmetry about the y-axis. Real circuits will have more complex power series expansions.

One common way of characterizing the linearity of a circuit is called the two-tone test. In this test, an input consisting of two sine waves is applied to the circuit:

$$v_{in} = v_1 \cos \omega_1 t + v_2 \cos \omega_2 t = X_1 + X_2 \tag{3.32}$$

When these tones are applied to the transfer function given in (3.29), the result is a number of terms:

$$v_0 = k_0 + \underbrace{k_1 (X_1 + X_2)}_{desired} + \underbrace{k_2 (X_1 + X_2)^2}_{second\,order} + \underbrace{k_3 (X_1 + X_2)^3}_{third\,order} \tag{3.33}$$

$$
\begin{aligned}
v_0 = k_0 &+ k_1 (X_1 + X_2) + k_2 \left(X_1^2 + 2X_1 X_2 + X_2^2 \right) \\
&+ k_3 \left(X_1^3 + 3X_1^2 X_2 + 3X_1 X_2^2 + X_1^3 \right)
\end{aligned}
\tag{3.34}
$$

These terms can be further broken down into various frequency components. For instance, the X_1^2 term has a component at dc and another at the second harmonic of the input:

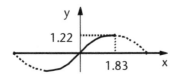

Figure 3.7 An example of a nonlinearity with first- and third-order terms.

$$X_1^2 = \left(v_1 \cos \omega_1 t\right)^2 = \frac{v_1^2}{2}\left(1 + \cos 2\omega_1 t\right) \tag{3.35}$$

The second-order terms can be expanded as follows:

$$\left(X_1 + X_2\right)^2 = \underbrace{X_1^2}_{\substack{dc \\ + \\ HD2}} + \underbrace{2X_1X_2}_{MIX} + \underbrace{X_2^2}_{\substack{dc \\ + \\ HD2}} \tag{3.36}$$

where second-order terms are comprised of second harmonics HD2, and mixing components, here labeled MIX, but sometimes labeled IM2 for second-order intermodulation. The mixing components will appear at the sum and difference frequencies of the two input signals. Note also that second-order terms cause an additional dc term to appear.

The third-order terms can be expanded as follows:

$$\left(X_1 + X_2\right)^3 = \underbrace{X_1^3}_{\substack{FUND \\ + \\ HD3}} + \underbrace{3X_1^2X_2}_{\substack{IM3 \\ + \\ FUND}} + \underbrace{3X_1X_2^2}_{\substack{IM3 \\ + \\ FUND}} + \underbrace{X_2^3}_{\substack{FUND \\ + \\ HD3}} \tag{3.37}$$

Third-order nonlinearity results in third harmonics HD3 and third-order intermodulation IM3. Expansion of both the HD3 terms and the IM3 terms show output signals appearing at the input frequencies. The effect is that third-order nonlinearity can change the gain, which is seen as gain compression. This is summarized in Table 3.2.

For completeness, fourth-order and fifth-order terms have also been shown in Table 3.2. Fourth-order terms are as follows:

$$\left(X_1 + X_2\right)^4 = X_1^4 + 4X_1^3X_2 + 6X_1^2X_2^2 + 4X_1X_2^3 + X_2^4 \tag{3.38}$$

As shown in Table 3.2 and illustrated in a later example, the fourth-order terms will add to the second order terms, that is, effecting the mixing components and the DC bias term.

Fifth-order terms, are as follows:

$$\left(X_1 + X_2\right)^5 = X_1^5 + 5X_1^4X_2 + 10X_1^3X_2^2 + 10X_1^2X_2^3 + 5X_1X_2^4 + X_2^5 \tag{3.39}$$

As shown in Table 3.2, and illustrated in a later example, the fifth-order terms will add to the distortion contributed by the third-order terms. As will be discussed later, third-order nonlinearity is often of critical importance in the design of a radio. Typically, such nonlinearity is dominated by the third-order terms and the fifth-order can be ignored; however, for higher input power, such as might be seen in a power amplifier or in cases where some cancellation scheme is used to cancel the third-order nonlinearity, the fifth-order effects can become very important.

Table 3.2 Summary of Distortion

Frequency	Component Amplitude for First-, Second-, and Third-Order Terms	Component Amplitude for Fourth- and Fifth-Order Terms
DC	$k_0 + \dfrac{k_2}{2}\left(v_1^2 + v_2^2\right)$	$k_4\left(\dfrac{3v_1^4}{8} + \dfrac{3v_1^2 v_2^2}{2} + \dfrac{3v_2^4}{8}\right)$
ω_1	$k_1 v_1 + k_3 v_1\left(\dfrac{3}{4}v_1^2 + \dfrac{3}{2}v_2^2\right)$	$k_5\left(\dfrac{5v_1^5}{8} + \dfrac{15v_1^3 v_2^2}{4} + \dfrac{15v_1 v_2^4}{8}\right)$
ω_2	$k_1 v_2 + k_3 v_2\left(\dfrac{3}{4}v_2^2 + \dfrac{3}{2}v_1^2\right)$	$k_5\left(\dfrac{15v_1^4 v_2}{8} + \dfrac{15v_1^2 v_2^3}{4} + \dfrac{5v_2^5}{8}\right)$
$2\omega_1$	$\dfrac{k_2 v_1^2}{2}$	$k_5\left(\dfrac{v_1^4}{2} + \dfrac{3v_1^2 v_2^2}{2}\right)$
$2\omega_2$	$\dfrac{k_2 v_2^2}{2}$	$k_4\left(\dfrac{3v_1^2 v_2^2}{2} + \dfrac{v_2^4}{2}\right)$
$\omega_2 \pm \omega_1$	$k_2 v_1 v_2$	$k_4\left(\dfrac{3v_1^3 v_2}{2} + \dfrac{3v_1 v_2^3}{2}\right)$
$3\omega_1$	$\dfrac{k_3 v_1^3}{4}$	$k_5\left(\dfrac{5v_1^5}{16} + \dfrac{5v_1^3 v_2^2}{4}\right)$
$3\omega_2$	$\dfrac{k_3 v_2^3}{4}$	$k_5\left(\dfrac{5v_1^2 v_2^3}{4} + \dfrac{5v_2^5}{16}\right)$
$2\omega_1 \pm \omega_2$	$\dfrac{3}{4}k_3 v_1^2 v_2$	$k_5\left(\dfrac{5v_1^4 v_2}{4} + \dfrac{15v_1^2 v_2^3}{8}\right)$
$2\omega_2 \pm \omega_1$	$\dfrac{3}{4}k_3 v_1 v_2^2$	$k_5\left(\dfrac{15v_1^3 v_2^2}{8} + \dfrac{5v_1 v_2^4}{4}\right)$
$2\omega_2 \pm 2\omega_1$	No term generated	$k_4\dfrac{3v_1^2 v_2^2}{4}$
$3\omega_1 \pm \omega_2$	No term generated	$k_4\dfrac{v_1^3 v_2^1}{2}$
$3\omega_2 \pm \omega_1$	No term generated	$k_4\dfrac{v_1^1 v_2^3}{2}$
$3\omega_1 \pm 2\omega_2$	No term generated	$k_5\dfrac{5v_1^3 v_2^2}{8}$
$3\omega_2 \pm 2\omega_1$	No term generated	$k_5\dfrac{5v_1^2 v_2^3}{8}$
$4\omega_1$	No term generated	$k_4\dfrac{v_1^4}{8}$
$4\omega_2$	No term generated	$k_4\dfrac{v_2^4}{8}$
$5\omega_1$	No term generated	$k_5\dfrac{v_1^5}{16}$
$5\omega_2$	No term generated	$k_5\dfrac{v_2^5}{16}$

Note that in the case of an amplifier, only the terms at the input frequency are desired. Of all the unwanted terms, the two at frequencies $2\omega_1 - \omega_2$ and $2\omega_2 - \omega_1$ are the most troublesome, as they can fall in the band of the desired outputs if ω_1 is close in frequency to ω_2 and therefore cannot be easily filtered out. These two tones are usually referred to as third-order intermodulation terms (IM3 products).

Example 3.4:   Determination of Frequency Components Generated in a Nonlinear System

Consider a nonlinear circuit with 7-MHz and 8-MHz tones applied at the input. Determine all output frequency components, assuming distortion components up to the fifth order.

Solution:
Table 3.3 and Figure 3.8 show the outputs.

It is apparent that harmonics can be filtered out easily, while the third-order intermodulation terms, being close to the desired tones, may be difficult to filter. Note that the fifth-order terms also add to the same intermodulation frequencies, however, while the third-order terms increase in amplitude at a rate of 3 dB for every 1 dB of input power, the tones originating from the fifth-order nonlinearity would increase by 5 dB for every 1 dB of increase of input power.

Table 3.3   Outputs from Nonlinear Circuits with Inputs at $f_1 = 7$, $f_2 = 8$ MHz

	Symbolic Frequency	Example Frequency	Name	Comment
First order	f_1, f_2	7, 8	Fundamental	Desired output
Second order	0	0	DC	Bias shifts
	$2f_1, 2f_2$	14, 16	HD2 (Harmonics)	Can filter
	$f_2 \pm f_1$	1, 15	IM2 (Mixing)	Can filter
Third order	f_1, f_2	7, 8	Fund	Cause compression
	$3f_1, 3f_2$	21, 24	HD3 (Harmonic)	Can filter harmonics
	$2f_1 \pm f_2$,	6, 22	IM3 (Intermod)	Difference frequencies close to fundamental,
	$2f_2 \pm f_1$	9, 23	IM3 (Intermod)	difficult to filter
Fourth order	0	0	DC	Bias shift
	$2f_1, 2f_2$	14, 16	HD2 (Harmonics)	Adds to second-order outputs (Can filter)
	$f_2 \pm f_1$	1, 15	IM2 (Mixing)	Adds to second-order outputs (Can filter)
	$4f_1, 4f_2$	28, 32	HD4	Can filter harmonics
	$3f_1 \pm f_2$	13, 29	IM4 (Mixing)	Can filter
	$3f_2 \pm f_1$	17, 31		
	$2f_2 \pm 2f_1$	2, 30	IM4 (Mixing)	Can filter
Fifth order	f_1, f_2	7, 8	Fund	Adds to third order outputs
	$3f_1, 3f_2$	21, 24	HD3 (Harmonics)	Adds to third order outputs
	$2f_1 \pm f_2$,	6, 22	IM3 (Intermod)	Same frequency components as for IP3,
	$2f_2 \pm f_1$	9, 23	IM3 (Intermod)	difficult to filter
	$5f_1, 5f_2$	35, 40	HD5 (Harmonics)	Can filter harmonics
	$4f_1 \pm f_2$	20, 36	IM5 (Intermod)	Can filter
	$4f_2 \pm f_1$	25, 39	IM5 (Intermod)	
	$3f_1 \pm 2f_2$	5, 37	IM5 (Intermod)	Difference frequencies close to fundamental,
			IM5 (Intermod)	difficult to filter
	$3f_2 \pm 2f_1$	10, 38	M5 (Intermod)	Difference frequencies close to fundamental,
			IM5 (Intermod)	difficult to filter

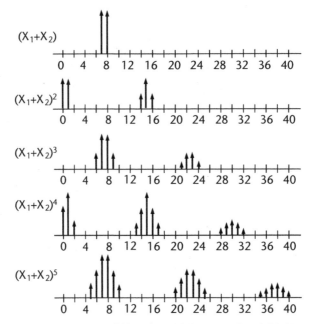

Figure 3.8 The output spectrum up to fifth order with inputs at 7 and 8 MHz.

3.4.2 Third-Order Intercept Point

One of the most common ways to test the linearity of a circuit is to apply two signals at the input, having equal amplitude and offset by some frequency, and plot fundamental output and intermodulation output power as a function of input power as shown in Figure 3.9. From the plot, the *third-order intercept point*

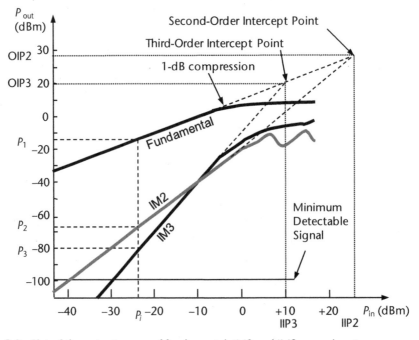

Figure 3.9 Plot of the output power of fundamental, IM3 and IM2 versus input power.

(IP3) is determined. The third-order intercept point is a theoretical point where the amplitudes of the intermodulation tones at $2\omega_1 - \omega_2$ and $2\omega_2 - \omega_1$ are equal to the amplitudes of the fundamental tones at ω_1 and ω_2.

From Table 3.2 if $v_1 = v_2 = v_i$, then the output voltage at the fundamental frequency is given by

$$\text{fund} = k_1 v_i + \frac{9}{4} k_3 v_i^3 \tag{3.40}$$

The linear component of (3.40) given by

$$\text{fund} = k_1 v_i \tag{3.41}$$

can be compared to the third-order intermodulation term given by

$$\text{IM3} = \frac{3}{4} k_3 v_i^3 \tag{3.42}$$

Note that for small v_i, the fundamental rises linearly (20 dB/decade) and that the IM3 terms rise as the cube of the input (60 dB/decade). A theoretical voltage at which these two tones will be equal can be defined:

$$\frac{3}{4} k_3 v_{\text{IP3}}^3 = k_1 v_{\text{IP3}} \tag{3.43}$$

This can be solved for v_{IP3}:

$$v_{\text{IP3}} = 2\sqrt{\frac{k_1}{3|k_3|}} \tag{3.44}$$

Note that for a circuit experiencing compression, k_3 must be negative. Thus, in (3.44) the absolute value of k_3 must be used. Note also that (3.44) gives the input voltage at the third-order intercept point. The input power at this point is called the *input third-order intercept point* (IIP3). If IP3 is specified at the output, it is called the *output third-order intercept point* (OIP3).

Of course, the third-order intercept point cannot actually be measured directly, because by the time the amplifier reached this point, it would be heavily overloaded. Therefore, it is useful to describe a quick way to extrapolate it at a given power level. Assume that a device with power gain G has been measured to have an output power of P_1 at the fundamental frequency and a power of P_3 at the IM3 frequency for a given input power of P_i, as illustrated in Figure 3.9. Now, on a log plot (for example, when power is in dBm) of P_3 and P_1 versus P_i, the IM3 terms have a slope of 3 and the fundamental terms have a slope of 1. Therefore,

$$\frac{\text{OIP3} - P_1}{\text{IIP3} - P_i} = 1 \tag{3.45}$$

$$\frac{\text{OIP3} - P_3}{\text{IIP3} - P_i} = 3 \tag{3.46}$$

because subtraction on a log scale amounts to division of power.

These equations can be solved to give

$$IIP3 = P_i + \frac{1}{2}[P_1 - P_3]$$ (3.47)

which also equals

$$IIP3 = P_1 + \frac{1}{2}[P_1 - P_3] - G$$ (3.48)

as

$$G = OIP3 - IIP3 = P_1 - P_i$$ (3.49)

3.4.3 Second-Order Intercept Point

A *second-order intercept point* (IP2) can be defined similar to that of the third-order intercept point. Which one is used depends largely on which is more important in the system of interest; for example, second-order distortion is particularly important in direct down conversion receivers. If two tones are present at the input, then the second-order output is given by

$$v_{IM2} = k_2 v_i^2$$ (3.50)

Note that the IM2 terms rise at 40 dB/dec rather than at 60 dB/dec, as in the case of the IM3 terms as shown in Figure 3.9.

The theoretical voltage at which the IM2 term will be equal to the fundamental term given in (3.41) can be defined:

$$k_2 v_{IP2}^2 = k_1 v_{IP2}$$ (3.51)

This can be solved for v_{IP2}:

$$v_{IP2} = \frac{k_1}{k_2}$$ (3.52)

Assume that a device with power gain G has been measured to have an output power of P_1 at the fundamental frequency and a power of P_2 at the IM2 frequency for a given input power of P_i, as illustrated in Figure 3.9. Now, on a log plot (for example, when power is in dBm) of P_2 and P_1 versus P_i, the IM2 terms have a slope of 2 and the fundamental terms have a slope of 1. Therefore:

$$\frac{OIP2 - P_1}{IIP2 - P_i} = 1$$ (3.53)

$$\frac{OIP2 - P_2}{IIP2 - P_i} = 2$$ (3.54)

These equations can be solved to give

$$IIP2 = P_i + [P_1 - P_2] = P_1 + [P_1 - P_2] - G$$ (3.55)

3.4.4   Fifth-Order Intercept Point

A *fifth-order intercept point* (IP5) can also be defined. If two tones are present the input at f_1 and f_2, each with an amplitude of v_i, the fifth-order outputs at $3f_2 \pm 2f_1$ and at $3f_1 \pm 2f_2$ are given by

$$v_{\mathrm{IM}\,5} = k_5 \frac{5v_i^5}{8}$$

(3.56)

Note that the IM5 terms rise at 100 dB/dec rather than at 60 dB/dec, as in the case of the IM3 terms.

The theoretical voltage at which the IM5 term will be equal to the fundamental term given in (3.41) can be defined:

$$k_5 \frac{5v_{\mathrm{IP5}}^5}{8} = k_1 v_{\mathrm{IP5}}$$

(3.57)

This can be solved for v_{IP5}:

$$v_{\mathrm{IP5}} = \left(\frac{8k_1}{5k_5} \right)^{1/4}$$

(3.58)

Assume that a device with power gain G has been measured to have an output power of P_1 at the fundamental frequency and a power of P_5 at the IM5 frequency for a given input power of P_i, as illustrated in Figure 3.9. Now, on a log plot (for example when power is in dBm) of P_5 and P_1 versus P_i, the IM5 terms have a slope of 5 and the fundamental terms have a slope of 1. Therefore

$$\frac{\mathrm{OIP5} - P_1}{\mathrm{IIP5} - P_i} = 1$$

(3.59)

$$\frac{\mathrm{OIP5} - P_5}{\mathrm{IIP5} - P_i} = 5$$

(3.60)

These equations can be solved to give

$$\mathrm{IIP5} = P_i + \frac{1}{4}\left[P_1 - P_5 \right] = P_1 + \frac{1}{4}\left[P_1 - P_5 \right] - G$$

(3.61)

3.4.5   The 1-dB Compression Point

In addition to measuring the IP3, IP2, or IIP5 of a circuit, the 1-dB compression point is another common way to measure linearity. This point is more directly measurable than IP3 and requires only one tone rather than two (although any number of tones can be used). The 1-dB compression point is simply the power level, specified at either the input or the output, where the output power is 1 dB less than it would have been in an ideally linear device. It is also marked in Figure 3.9.

We first note that at 1-dB compression, the ratio of the actual output voltage v_o to the ideal output voltage v_{oi} is

$$20\log_{10}\left(\frac{v_o}{v_{oi}}\right) = -1\,\text{dB}$$

(3.62)

or

$$\frac{v_o}{v_{oi}} = 0.89125$$

(3.63)

Now referring again to Table 3.2, it is noted that the actual output voltage for a single tone is

$$v_o = k_1 v_i + \frac{3}{4} k_3 v_i^3$$

(3.64)

for an input voltage v_i. The ideal output voltage is given by

$$v_{oi} = k_1 v_i$$

(3.65)

Thus, the 1-dB compression point can be found by substituting (3.64) and (3.65) into (3.63):

$$\frac{k_1 v_{1\text{dB}} + \frac{3}{4} k_3 v_{1\text{dB}}^3}{k_1 v_{1\text{dB}}} = 0.89125$$

(3.66)

Note that for a nonlinearity that causes compression, rather than one that causes expansion, k_3 has to be negative. Solving (3.66) for $v_{1\text{dB}}$ gives

$$v_{1\text{dB}} = 0.38\sqrt{\frac{k_1}{|k_3|}}$$

(3.67)

If more than one tone is applied, the 1-dB compression point will occur for a lower input voltage. In the case of two equal amplitude tones applied to the system, the actual output power for one frequency is

$$v_o = k_1 v_i + \frac{9}{4} k_3 v_i^3$$

(3.68)

The ideal output voltage is still given by (3.65). So now the ratio is

$$\frac{k_1 v_{1\text{dB}} + \frac{9}{4} k_3 v_{1\text{dB}}^3}{k_1 v_{1\text{dB}}} = 0.89125$$

(3.69)

Therefore, the 1-dB compression voltage is now

$$v_{1\text{dB}} = 0.22\sqrt{\frac{k_1}{|k_3|}}$$

(3.70)

Thus, as more tones are added, this voltage will continue to get lower.

3.4.6   Relationships Between 1-dB Compression and IP3 Points

In previous sections, formulas for the IP3 and the 1-dB compression point have been derived. As we now have expressions for both these values, we can find a relationship between these two points. Taking the ratio of (3.44) and (3.67) gives

$$\frac{v_{IP3}}{v_{1dB}} = \frac{2\sqrt{\dfrac{k_1}{3|k_3|}}}{0.38\sqrt{\dfrac{k_1}{|k_3|}}} = 3.04 \tag{3.71}$$

Thus, these voltages are related by a factor of 3.04 or about 9.66 dB, independent of the particulars of the nonlinearity in question. In the case of the 1-dB compression point with two tones applied, the ratio is larger. In this case

$$\frac{v_{IP3}}{v_{1dB}} = \frac{2\sqrt{\dfrac{k_1}{3|k_3|}}}{0.22\sqrt{\dfrac{k_1}{|k_3|}}} = 5.25 \tag{3.72}$$

These voltages are related by a factor of 5.25 or about 14.4 dB.

Thus, one can estimate that for a single tone, the compression point is about 10 dB below the intercept point, while for two tones; the 1-dB compression point is close to 15 dB below the intercept point. The difference between these two numbers is just the factor of three (4.77 dB) resulting from the second tone.

Note that this analysis is valid for third-order nonlinearity. For stronger non-linearity (i.e., containing fifth-order terms), additional components are found at the fundamental as well as at the intermodulation frequencies. Care must also be taken here as this analysis also assumes that there is only one nonlinearity in the system, and if there are two or more, this relationship will likely not hold true. Also note that for two-tone analysis to be valid, the frequency spacing must be narrow relative to the circuit bandwidth.

Example 3.5:   Determining IIP3 and 1-dB Compression Point from Measurement Data

An amplifier designed to operate at 2 GHz with a gain of 10 dB has two signals of equal power applied at the input. One is at a frequency of 2.0 GHz and another is at a frequency of 2.01 GHz. At the output, four tones are observed at 1.99, 2.00, 2.01, and 2.02 GHz. The power levels of the tones are −70, −20, −20, and −70 dBm, respectively. Determine the IIP3 and 1-dB compression point for this amplifier.

Solution:
The tones at 1.99 and 2.02 GHz are the IP3 tones. We can use (3.47) directly to find the IIP3

$$IIP3 = P_1 + \frac{1}{2}\left[P_1 - P_3\right] - G = -20 + \frac{1}{2}\left[-20 + 70\right] - 10 = -5 \text{ dBm}$$

The 1-dB compression point for a single tone is 9.66 dB lower than this value, about −14.7 dBm at the input.

3.4.7 Broadband Measures of Linearity

Intercept points and 1-dB compression point are two common measures of linearity, but they are by no means the only ones. Many others exist and, in fact, more could be defined as is convenient. Two other measures of linearity that are common in wideband systems handling many signals simultaneously are called *composite triple-order beat* (CTB) and *composite second-order beat* (CSO) [10, 11]. These definitions of linearity can also be very helpful in other communication systems that use OFDM modulation such as wireless LAN or ultrawideband radios. In these tests of linearity, N signals of voltage v_i are applied to the circuit equally spaced in frequency, as shown in Figure 3.10. Note here that, as an example, the tones are spaced 6 MHz apart (this is the spacing for a cable television system for which this is a popular way to characterize linearity). Note also that the tones are never placed at a frequency that is an exact multiple of the spacing (in this case, 6 MHz). This is done so that third-order terms and second-order terms fall at different frequencies. This will be clarified shortly.

If we take three of these signals, then the third-order nonlinearity gets a little more complicated than before:

$$(x_1 + x_2 + x_3)^3 = \underbrace{x_1^3 + x_2^3 + x_3^3}_{\text{HM3}} + \underbrace{3x_1^2 x_2 + 3x_1^2 x_3 + 3x_2^2 x_1 + 3x_3^2 x_1 + 3x_2^2 x_3 + 3x_3^2 x_2}_{\text{IM3}}$$

$$+ \underbrace{6x_1 x_2 x_3}_{\text{TB}} \tag{3.73}$$

The last term in the expression causes CTB, in that it creates terms at frequencies $\omega_1 \pm \omega_2 \pm \omega_3$ of magnitude $1.5k_3 v_i$ where $\omega_1 < \omega_2 < \omega_3$. The TB term is twice as large as the IM3 products. Note that, except for the case where all three add ($\omega_1 + \omega_2 + \omega_3$), these tones can fall into any of the channels being used and many will fall into the same channel. For instance, in Figure 3.10, $67.25 − 73.25 + 79.25 = 73.25$ MHz, or

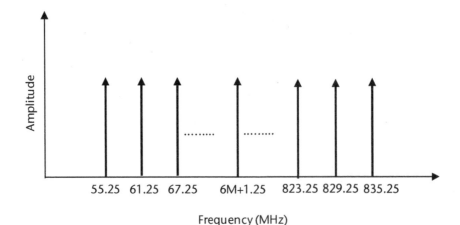

Frequency (MHz)

Figure 3.10 Equally spaced tones entering a broadband circuit.

49.25 – 55.25 + 79.25 = 73.25 MHz will both fall on the 73.25-MHz frequency. In fact, there will be many more *triple beat* (TB) products than IM3 products. Thus, these terms become more important in a wideband system. It can be shown that the maximum number of terms will fall on the tone at the middle of the band. With N tones, it can also be shown that the number of tones falling there will be

$$\text{Tones} = \frac{3}{8} N^2$$

(3.74)

We have already said that the voltage of these tones is twice that of the IM3 tones. It is also noted here that if the signal power is backed off from the IP3 power by some amount, then the power in the IM3 tones will be backed off three times as much (calculated on a logarithmic scale). Therefore, if each fundamental tone is at a power level of P_s (in dBm), then the power of the TB tones will be

$$\text{TB}(\text{dBm}) = P_{\text{IP3}} - 3\left(P_{\text{IP3}} - P_s\right) + 6$$

(3.75)

where P_{IP3} is the IP3 power level for the given circuit.

Now, assuming that all tones add as power rather than voltage, and noting that CTB is usually specified as so many dB down from the signal power,

$$\text{CTB}(\text{dB}) = P_s - \left[P_{\text{IP3}} - 3\left(P_{\text{IP3}} - P_s\right) + 6 + 10\log\left(\frac{3}{8} N^2\right) \right]$$

(3.76)

Note that CTB could be found using either input or output referred power levels.

Similar to the CTB is the CSO, which can also be used to measure the linearity of a broadband system. Again, if we have N signals all at the same power level, we now consider the second-order distortion products of each pair of signals that fall at frequencies $\omega_1 \pm \omega_2$. In this case, the signals fall at frequencies either above or below the carriers rather than right on top of them, as in the case of the triple beat terms, provided that the carriers are not some even multiple of the channel spacing. For example, in Figure 3.10, 49.25 + 55.25 = 104.5 MHz is 1.25 MHz above the closest carrier at 103.25 MHz. All the sum terms will fall 1.25 MHz above the closest carrier, while the difference terms such as 763.25 – 841.25 = 78 MHz will fall 1.25 MHz below the closest carrier at 79.25 MHz. Thus, the second-order and third-order terms can be measured separately. The number of terms that fall next to any given carrier will vary. Some of the $\omega_1 + \omega_2$ terms will fall out of band and the maximum number in band will fall next to the highest frequency carrier. The number of second-order beats above any given carrier is given by

$$N_B = (N-1)\frac{f - 2f_L + d}{2(f_H - f_L)}$$

(3.77)

where N is the number of carriers, f is the frequency of the measurement channel, f_L is the frequency of the lowest channel, f_H is the frequency of the highest channel, and d is the frequency offset from a multiple of the channel spacing (1.25 MHz in Figure 3.10).

For the case of the difference frequency second-order beats, there are more of these at lower frequencies, and the maximum number will be next to the lowest frequency carrier. In this case, the number of second-order products next to any carrier can be approximated by

$$N_B = (N-1)\left(1 - \frac{f-d}{f_H - f_L}\right)$$

(3.78)

Each of the second-order beats is an IP2 tone. Therefore, if each fundamental tone is at a power level of P_s, then the power of the *second-order beat* (SO) tones will be

$$SO\,(dBm) = P_{IP2} - 2(P_{IP2} - P_s)$$

(3.79)

Thus, the composite second-order beat product will be given by

$$CSO(dB) = P_s - \left[P_{IP2} - 2(P_{IP2} - P_s) + 10\log(N_B)\right]$$

(3.80)

Obviously, the three-tone example with third-order nonlinearity dealt with above can be extended to higher-order nonlinearity, for example, fourth and fifth order, which will produce additional higher-order terms but will also produce output tones similar to that produced by the third-order nonlinearity. This was seen earlier in Section 3.4.1, Table 3.2, and Figure 3.8 for two inputs with nonlinearity up to the fifth order. This sort of analysis can also be extended to have inputs with more than three tones, for example, four tones or five tones produce more beats (quarto beats, penta beats) and obviously very long expressions. Generally, if a nonlinearity cannot be modeled with a third-order series, it is best to employ a simulation or numerical technique.

3.4.8 Nonlinearity with Feedback

Consider a system with feedback as shown in Figure 3.11. Assume also that in this system that the amplifier with gain A is nonlinear with a nonlinearity that can be described by a third-order power series. We would like to determine the nonlinearity for the overall system in closed loop. To do this, we start by stating that the error signal is given by

$$v_{error} = v_{in} - v_{fb} = v_{in} - B v_{out}$$

(3.81)

As v_{error} is the input to the nonlinear gain stage A, the output of A will be given by

$$v_{out} = k_1 v_{error} + k_2 v_{error}^2 + k_3 v_{error}^3$$

(3.82)

Now substituting (3.81) into (3.82) gives

$$v_{out} = k_1\left(v_{in} - B v_{out}\right) + k_2\left(v_{in} - B v_{out}\right)^2 + k_3\left(v_{in} - B v_{out}\right)^3$$

(3.83)

which can be expanded to

$$v_{out} = k_1 \left(v_{in} - B v_{out} \right)$$

$$+ k_2 \left(v_{in}^2 - 2B v_{in} v_{out} + B^2 v_{out}^2 \right) \tag{3.84}$$

$$+ k_3 \left(v_{in}^3 - 3B v_{in}^2 v_{out} + 3B^2 v_{in} v_{out}^2 - B^3 v_{out}^3 \right)$$

Assuming that the overall nonlinearity will be given by

$$v_{out} = g_1 v_{in} + g_2 v_{in}^2 + g_3 v_{in}^3 \tag{3.85}$$

Equation (3.85) can be substituted into (3.83) to obtain

$$v_{out} = k_1 \left(v_{in} - B \left(g_1 v_{in} + g_2 v_{in}^2 + g_3 v_{in}^3 \right) \right)$$

$$+ k_2 \left(v_{in}^2 - 2B v_{in} \left(g_1 v_{in} + g_2 v_{in}^2 + g_3 v_{in}^3 \right) + B^2 \left(g_1 v_{in} + g_2 v_{in}^2 + g_3 v_{in}^3 \right)^2 \right)$$

$$+ k_3 \left(\begin{array}{l} v_{in}^3 - 3B v_{in}^2 \left(g_1 v_{in} + g_2 v_{in}^2 + g_3 v_{in}^3 \right) + 3B^2 v_{in} \left(g_1 v_{in} + g_2 v_{in}^2 + g_3 v_{in}^3 \right)^2 \\ - B^3 \left(g_1 v_{in} + g_2 v_{in}^2 + g_3 v_{in}^3 \right)^3 \end{array} \right) \tag{3.86}$$

Looking at only the first-order terms (they must total to $g_1 v_{in}$):

$$g_1 v_{in} = k_1 v_{in} - B k_1 g_1 v_{in} \tag{3.87}$$

and solving for g_1 gives

$$g_1 = \frac{k_1}{1 + k_1 B} \tag{3.88}$$

Looking at the second-order terms:

$$g_2 v_{in}^2 = -k_1 B g_2 v_{in}^2 + k_2 \left(v_{in} - B g_1 v_{in} \right)^2 \tag{3.89}$$

and solving for g_2 gives

$$g_2 = \frac{k_2 \left(1 - B g_1 \right)^2}{1 + k_1 B} \tag{3.90}$$

Substituting in for g_1 gives final result of

$$g_2 = \frac{k_2}{\left(1 + k_1 B \right)^3} \tag{3.91}$$

Now looking at the third-order terms:

$$g_3 v_{in}^3 = -k_1 B g_3 v_{in}^3 - 2 k_2 B g_2 \left(1 - B g_1 \right) v_{in}^3 + k_3 \left(v_{in} - B g_1 v_{in} \right)^3 \tag{3.92}$$

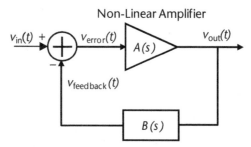

Figure 3.11 A feedback system that includes a nonlinear amplifier.

and solving for g_3 gives

$$g_3 = \frac{k_3 \left(1 - Bg_1\right)^3 - 2Bk_2 g_2 \left(1 - Bg_1\right)}{1 + k_1 B}$$

(3.93)

Substituting in for g_1 and g_2 gives

$$g_3 = \frac{k_3 \left(1 + Bk_1\right) - 2Bk_2^2}{\left(1 + k_1 B\right)^5}$$

(3.94)

Note that from this derivation that even a system that includes a device with only a second-order nonlinearity will produce an overall system with a third-order term that is nonzero. Also note that if

$$B = \frac{k_3}{2k_2^2 + k_1 k_3}$$

(3.95)

the third-order term can be made to go to zero, thus removing third-order nonlinearities from the system.

Example 3.6:   Amplifier with Resistive Emitter Degeneration

Use the theory developed in this section to derive the v_{ip3} of a common-emitter bipolar amplifier with resistive degeneration shown in Figure 3.12. Note that the base-emitter voltage to collector current characteristic is well known and given by

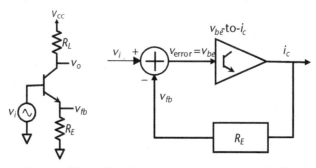

Figure 3.12 A transistor amplifier with resistive degeneration and a feedback model to describe the nonlinearity.

$$I_C + i_c = I_S e^{\frac{V_{BE}}{v_T}} e^{\frac{V_{be}}{v_T}} = I_C e^{\frac{V_{be}}{v_T}} = \frac{I_C}{v_T} v_{be} + \frac{I_C}{2v_T^2} v_{be}^2 + \frac{I_c}{6v_T^3} v_{be}^3 + \ldots$$

Solution:
In this case, the circuit can be mapped to the feedback system by noting that the transistor converts the voltage into collector current. The emitter current (which is approximately equal to the collector current) creates the feedback voltage by passing that same current across the feedback resistor R_E. Thus, it should be obvious how to map the circuit to the model in this case. After doing this, we note the value of the feedforward coefficients directly from the power series for the transistor:

$$k_1 = \frac{I_C}{v_T} = \frac{1}{r_e}$$

$$k_2 = \frac{I_C}{2v_T^2} = \frac{1}{2I_C r_e^2}$$

$$k_3 = \frac{I_C}{6v_T^3} = \frac{1}{6I_C^2 r_e^3}$$

Now we can find the overall nonlinearity for the complete amplifier making use of the results in this section:

$$g_1 = \frac{k_1}{1 + k_1 B(s)} = \frac{1}{r_e + R_E}$$

$$g_2 = \frac{k_2}{\left(1 + k_1 B(s)\right)^3} = \frac{r_e}{2I_C \left(r_e + R_E\right)^3}$$

$$g_3 = \frac{k_3}{\left(1 + k_1 B(s)\right)^4} - \frac{2B(s)k_2^2}{\left(1 + k_1 B(s)\right)^5}$$

$$= \left[\frac{1}{6I_C^2}\left(\frac{r_e}{r_e + R_E}\right) - \frac{R_E}{2r_e I_C^2}\left(\frac{r_e}{r_e + R_E}\right)^2 \right]\left[\frac{1}{r_e + R_E} \right]^3 = \frac{r_e}{6I_C^2}\frac{r_e - 2R_E}{\left(r_e + R_E\right)^5}$$

As a result, v_{IP3} can be determined as:

$$v_{IP3} = 2\sqrt{\frac{g_1}{3g_3}} = 2\sqrt{\frac{6I_C^2 \left(r_e + R_E\right)^5}{3r_e \left(r_e + R_E\right)\left(r_e - 2R_E\right)}} = 2\sqrt{2}v_T \frac{\left(r_e + R_E\right)^2}{\sqrt{r_e^3 \left(r_e - 2R_E\right)}}$$

3.4.9 Nonlinear Systems with Memory: Volterra Series

A simple power series, as discussed in the previous sections, cannot handle a system with memory. Memory, for example, due to inductors and capacitors, will result in systems where the response is frequency dependent. Although, in general, amplifiers and many other circuits used in communications applications make use of inductors and capacitors for tuning, matching, or filtering, often simple power series are used because of their inherent simplicity. However, to describe the system more accurately, other techniques are sometimes used, for example, using Volterra series. It should be noted that Volterra series analysis, like simple power series analysis is most useful in systems with only mild nonlinearity.

In this section, we will start with the simplest way to combine nonlinearity with memory, which is to start with a memoryless nonlinear circuit then add circuits (filters) with memory at the input and the output. As a starting point, consider a memoryless nonlinear circuit as shown in Figure 3.13. As discussed in an earlier section, the input-output relationship is described by

$$v_{out} = k_1 v_{in} + k_2 v_{in}^2 + k_3 v_{in}^3 \tag{3.96}$$

Here the nonlinearities are simply drawn out graphically using multipliers and adders. This is helpful to see how the signals pass through the nonlinearity, which is more important when dealing with systems with memory.

Now we will deal with a system that has a frequency dependent network attached to the input and the output of the nonlinear device (this could be an LRC network). This is shown in Figure 3.14. This is straightforward if the input $V_{in}(t)$ is described by

$$v_{in} = v_1 \cos \omega_1 t + v_1 \cos \omega_2 t \tag{3.97}$$

Then $V_{in}'(t)$ is given by

$$V_{in}'(t) = v_1 T_{in}(j\omega_1)\cos \omega_1 t + v_1 T_{in}(j\omega_2)\cos \omega_2 t \tag{3.98}$$

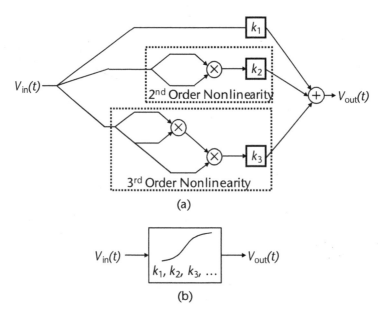

(a)

(b)

Figure 3.13 (a) Representation of a memoryless nonlinear block showing nonlinearity up to third order, and (b) simplified diagram of a memoryless nonlinear block.

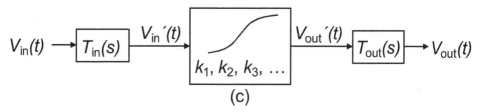

Figure 3.14   A system with input and output blocks that have memory but are linear and in between is a memory-less nonlinear block.

Table 3.4   Summary of Distortion Components with Input and Output Memory Circuits

ω	T_{in}	Nonlinearity	T_{out}	Order Source	Description
dc	$T_{in}^2(j\omega_1)$	$\dfrac{k_2 v_1^2}{2}$	$T_{out}(0)$	Second	DC offset terms
dc	$T_{in}^2(j\omega_1)$	$\dfrac{k_2 v_1^2}{2}$	$T_{out}(0)$	Second	DC offset terms
ω_1	$T_{in}(j\omega_2)$	$k_1 v_1$	$T_{out}(j\omega_1)$	First	Desired linear output
ω_1	$T_{in}^3(j\omega_1)$	$\dfrac{3}{4}k_3 v_1^3$	$T_{out}(j\omega_1)$	Third	Distortion at desired output frequency
ω_1	$T_{in}(j\omega_1)T_{in}^2(j\omega_2)$	$\dfrac{3}{2}k_3 v_1^3$			
ω_2	$T_{in}(j\omega_2)$	$k_1 v_1$	$T_{out}(j\omega_2)$	First	Desired linear output
ω_2	$T_{in}^3(j\omega_2)$	$\dfrac{3}{4}k_3 v_1^3$	$T_{out}(j\omega_2)$	Third	Distortion at desired output frequency
ω_2	$T_{in}^2(j\omega_1)T_{in}(j\omega_2)$	$\dfrac{3}{2}k_3 v_1^3$			
$2\omega_1$	$T_{in}^2(j\omega_1)$	$\dfrac{1}{2}k_2 v_1^2$	$T_{out}(j2\omega_1)$	Second	Second harmonic
$2\omega_2$	$T_{in}^2(j\omega_2)$	$\dfrac{1}{2}k_2 v_1^2$	$T_{out}(j2\omega_2)$	Second	Second harmonic
$\omega_2-\omega_1$	$T_{in}(j\omega_1)T_{in}(j\omega_2)$	$k_2 v_1^2$	$T_{out}(j(\omega_2-\omega_1))$	Second	Mixing term
$\omega_2+\omega_1$	$T_{in}(j\omega_1)T_{in}(j\omega_2)$	$k_2 v_1^2$	$T_{out}(j(\omega_2-\omega_1))$	Second	Mixing term
$3\omega_1$	$T_{in}^3(j\omega_1)$	$\dfrac{1}{4}k_3 v_1^3$	$T_{out}(j3\omega_1)$	Third	Third harmonic
$3\omega_2$	$T_{in}^3(j\omega_2)$	$\dfrac{1}{4}k_3 v_1^3$	$T_{out}(j3\omega_2)$	Third	Third harmonic
$2\omega_1-\omega_2$	$T_{in}^2(j\omega_1)T_{in}(j\omega_2)$	$\dfrac{3}{4}k_3 v_1^3$	$T_{out}(j(2\omega_1-\omega_2))$	Third	Third-order intermodulation
$2\omega_2-\omega_1$	$T_{in}(j\omega_1)T_{in}^2(j\omega_2)$	$\dfrac{3}{4}k_3 v_1^3$	$T_{out}(j(2\omega_2-\omega_1))$	Third	Third-order intermodulation
$2\omega_1+\omega_2$	$T_{in}^2(j\omega_1)T_{in}(j\omega_2)$	$\dfrac{3}{4}k_3 v_1^3$	$T_{out}(j(2\omega_1-\omega_2))$	Third	Third-order intermodulation
$2\omega_2+\omega_1$	$T_{in}(j\omega_1)T_{in}^2(j\omega_2)$	$\dfrac{3}{4}k_3 v_1^3$	$T_{out}(j(2\omega_2-\omega_1))$	Third	Third-order intermodulation

Table 3.5 Simplified Summary of Distortion Components with Input and Output Memory Circuits

ω	T_{in}	Nonlinearity	T_{out}	Order Source	Description
dc	$T_{in}^2(j\omega_0)$	$k_2 v_1^2$	$T_{out}(0)$	Second	dc offset terms
ω_1	$T_{in}^3(j\omega_0)$	$k_1 v_1$	$T_{out}(j\omega_0)$	First	Desired linear output
ω_1	$T_{in}^3(j\omega_0)$	$\dfrac{9}{4}k_3 v_1^3$	$T_{out}(j\omega_0)$	Third	Distortion at desired output frequency
ω_2	$T_{in}(j\omega_0)$	$k_1 v_1$	$T_{out}(j\omega_0)$	First	Desired linear output
ω_2	$T_{in}^3(j\omega_0)$	$\dfrac{9}{4}k_3 v_1^3$	$T_{out}(j\omega_0)$	Third	Distortion at desired output frequency
$2\omega_1$	$T_{in}^2(j\omega_0)$	$\dfrac{1}{2}k_2 v_1^2$	$T_{out}(j2\omega_0)$	Second	Second harmonic
$2\omega_2$	$T_{in}^2(j\omega_0)$	$\dfrac{1}{2}k_2 v_1^2$	$T_{out}(j2\omega_0)$	Second	Second harmonic
$\omega_2 - \omega_1$	$T_{in}^2(j\omega_0)$	$k_2 v_1^2$	$T_{out}(j\Delta\omega)$	Second	Mixing term
$\omega_2 + \omega_1$	$T_{in}^2(j\omega_0)$	$k_2 v_1^2$	$T_{out}(j2\omega_0)$	Second	Mixing term
$3\omega_1$	$T_{in}^3(j\omega_0)$	$\dfrac{1}{4}k_3 v_1^3$	$T_{out}(j3\omega_0)$	Third	Third harmonic
$3\omega_2$	$T_{in}^3(j\omega_0)$	$\dfrac{1}{4}k_3 v_1^3$	$T_{out}(j3\omega_0)$	Third	Third harmonic
$2\omega_1 - \omega_2$	$T_{in}^3(j\omega_0)$	$\dfrac{3}{4}k_3 v_1^3$	$T_{out}(j\omega_0)$	Third	Third-order intermodulation
$2\omega_2 - \omega_1$	$T_{in}^3(j\omega_0)$	$\dfrac{3}{4}k_3 v_1^3$	$T_{out}(j\omega_0)$	Third	Third-order intermodulation
$2\omega_1 + \omega_2$	$T_{in}^3(j\omega_0)$	$\dfrac{3}{4}k_3 v_1^3$	$T_{out}(j3\omega_0)$	Third	Third-order intermodulation
$2\omega_2 + \omega_1$	$T_{in}^3(j\omega_0)$	$\dfrac{3}{4}k_3 v_1^3$	$T_{out}(j3\omega_0)$	Third	Third-order intermodulation

That is, each V_{in} component is passed through the input memory system and results in a V'_{in} component at the same frequency but with potentially altered in magnitude or phase. These V'_{in} components are then passed through the nonlinearity, generating a bunch of different V'_{out} components at different frequencies. For a nonlinearity of order n, these frequencies will be sum and difference frequencies of all harmonics of the input frequencies up to the nth harmonic. Each of these is then passed through the output memory components, requiring that each V'_{out} component be multiplied by T_{out} at the appropriate frequency. Thus, Table 3.2 can be repeated here with the appropriate values of T_{in} and T_{out} added to it and is shown in Table 3.4.

Now let us consider some useful simplifications. First, let us assume that we have applied narrowband inputs to the system so that ω_1 and ω_2 are close in frequency so that there is not much phase or amplitude shift in either T_{in} or T_{out} between these two frequencies. So let us assume that

$$\omega_0 = \frac{\omega_1 + \omega_2}{2} \approx \omega_1 \approx \omega_2 \approx 2\omega_2 - \omega_1 \approx 2\omega_1 - \omega_2$$

$$(3.99)$$

Also let us assume that

$$2\omega_1 \approx 2\omega_2 \approx \omega_2 + \omega_1 \approx 2\omega_0 \tag{3.100}$$

Often, we can also assume that

$$\omega_2 - \omega_1 = \Delta\omega \tag{3.101}$$

And finally:

$$3\omega_2 \approx 3\omega_1 \approx 2\omega_1 + \omega_2 \approx 2\omega_2 + \omega_1 \approx 3\omega_0 \tag{3.102}$$

As a result, Table 3.4 simplifies to Table 3.5.

Nonlinear circuits with feedback will now be studied with reference to the block diagram in Figure 3.15. It is assumed here that the feed forward block is a memory-less nonlinear system and the feedback block is a linear system with memory. We will be analyzing this as a weakly nonlinear circuit. This means the input power will be kept sufficiently low such that the distortion components are much less than the fundamental component, and there will be no visible gain compression. Hence, the only important fundamental tone will be the desired one. As the main goal in this analysis will be to determine third-order intermodulation, this assumption is completely valid. This can also be shown with the full details in Figure 3.16.

In a system without feedback, the input signal going through the third-order block k_3 produces third-order output. In a system with feedback, a second mechanism is due to second-order output components being fed back and mixing with the fundamental input components to generate third-order outputs. To determine the third-order outputs, due to both mechanisms, the steps are:

1. Determine the fundamental output v_{out1} due to the applied input signals going through the linear component k_1.
2. Determine the error component v_{err1} due to the applied input signals and the linear block k_1.
3. Apply v_{err1} to the second-order nonlinearity k_2 generating dc and second harmonic v_{out2}.
4. Feedback of v_{out2} through B produces v_{err2} that is mixed with v_{err1} as it passes through the second-order block. This generates a third-order term $v_{out3}{}'$.

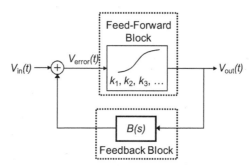

Figure 3.15   A nonlinear feedforward block with feedback.

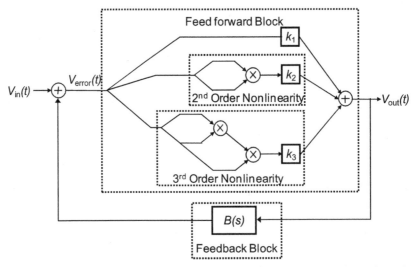

Figure 3.16 A nonlinear feedforward block with feedback showing each nonlinear branch.

5. Apply v_{err1} to the third-order nonlinearity k_3 generating a fundamental and third-order term v_{out3}''. This fundamental is not considered further due to the assumption of weak nonlinearity.

Note that for each output term, we need to include the effect of the feedback loop consisting of $B(s)$ and the linear term k_1.

To illustrate the above process, this analysis will first be performed for a feedback block $B(s)$ that does not have any frequency dependence, for example, a resistive network. The result can be verified by comparing to a previously derived result. Assume once more that we have two input signals:

$$v_{\text{in}} = v\cos\omega_1 t + v\cos\omega_2 t \qquad (3.103)$$

For step (1), if this signal passes through the linear part of the feedforward block, it will have an input to output transfer function of forward gain divided by one plus the loop gain, or:

$$\frac{v_{\text{out}}}{v_{\text{in}}} = \frac{k_1}{1+k_1B} \qquad (3.104)$$

Therefore, we can write

$$v_{\text{out1}}\left(t\right) = \frac{k_1}{1+k_1B}\left[v\cos\omega_1 t + v\cos\omega_2 t\right] \qquad (3.105)$$

In step (2), these signals produce an error voltage:

$$v_{\text{err1}} = \frac{1}{1+k_1B}\left[v\cos\omega_1 t + v\cos\omega_2 t\right] \qquad (3.106)$$

As noted earlier, other tones at the fundamental frequency will appear due to nonlinearity, but these are not important until closer to the compression point.

In step (3), v_{err1} is also applied to the second-order nonlinearity part of the feedforward system. The output is:

$$v_{out2}(t) = \left\{\frac{1}{1+k_1 B}[v\cos\omega_1 t + v\cos\omega_2 t]\right\}^2 \frac{k_2}{1+k_1 B} \tag{3.107}$$

Note that this second-order output signal will experience the feedback loop consisting of B and linear feedforward component k_1; thus, this signal also is attenuated by a factor of one plus the loop gain. The overall function can be written as

$$v_{out2}(t) = \frac{k_2}{(1+k_1 B)^3}[v\cos\omega_1 t + v\cos\omega_2 t]^2 \tag{3.108}$$

In step (4), v_{out2} is multiplied by $B(s)$ and fed back as error signal v_{err2} and along with v_{err1} is passed through the second-order term generating an output of $k_2(v_{err1} + v_{err2})^2$. Although this also produces dc terms and second harmonics, we will concentrate on the mixing term of $2k_2(v_{err1} \times v_{err2})$ to produce third-order terms at the output. This output is also attenuated by one plus the loop gain (the negative sign is due to the fact that the signal in fed into the negative input terminal):

$$v_{out3}{}'(t) = 2\left\{\frac{-Bk_2}{(1+k_1 B)^3}[v\cos\omega_1 t + v\cos\omega_2 t]^2\right\} \times \left\{\frac{1}{(1+k_1 B)}[v\cos\omega_1 t + v\cos\omega_2 t]\right\}$$

$$\times \frac{k_2}{(1+k_1 B)} = \frac{-2Bk_2^2}{(1+k_1 B)^5}[v\cos\omega_1 t + v\cos\omega_2 t]^3 \tag{3.109}$$

In step (5), the fundamental error signal, v_{err1} signal also passes through the third-order nonlinearity producing

$$v_{out3}{}''(t) = \frac{k_3}{(1+k_1 B)^3}[v\cos\omega_1 t + v\cos\omega_2 t]^3 \frac{1}{(1+k_1 B)} \tag{3.110}$$

Therefore, the overall third-order term is

$$v_{out3}(t) = \left(\frac{k_3}{(1+k_1 B)^4} - \frac{2Bk_2^2}{(1+k_1 B)^5}\right)[v\cos\omega_1 t + v\cos\omega_2 t]^3 \tag{3.111}$$

This agrees with (3.94).

The next added complexity is to assume that $B(s)$ has some frequency dependence and repeat the analysis. As in the previous example, most of the results are multiplied by the inverse of one plus the loop gain, so to simplify the expressions, this will be replaced by $T(s)$ as follows:

$$T(s) = \frac{1}{1+k_1 B(s)} \tag{3.112}$$

In step (1), v_{out1} is given by

$$v_{out1}(t) = k_1 T(j\omega_1) v \cos\omega_1 t + k_1 T(j\omega_2) v \cos\omega_2 t \qquad (3.113)$$

In step (2), the error voltage v_{err1} due to the fundamental term is

$$v_{err1} = T(j\omega_1) v \cos\omega_1 t + T(j\omega_2) v \cos\omega_2 \qquad (3.114)$$

In step (3), passing v_{err1} through the second-order nonlinearity results in

$$v_{out2} = \frac{v^2 k_2}{2}\begin{bmatrix} T^2(j\omega_1)T(j2\omega_1)\cos 2\omega_1 t \\ + T^2(j\omega_2)T(j2\omega_2)\cos 2\omega_2 t \\ + T^2(j\omega_1)T(0) \\ + T^2(j\omega_2)T(0) \\ + 2T(j\omega_1)T(j\omega_2)T(j(\omega_1+\omega_2))\cos(\omega_1+\omega_2)t \\ + 2T(j\omega_1)T(j\omega_2)T(j(\omega_1-\omega_2))\cos(\omega_1-\omega_2)t \end{bmatrix} \qquad (3.115)$$

In step (4), the output $v_{out2}(t)$ is fed back through $B(s)$, becoming v_{err2} and is then multiplied with $v_{err1}(t)$ as follows:

$$v_3' = v_{err2} \times v_{err1} =$$

$$\frac{-v^3 k_2^2}{2}\begin{bmatrix} T^2(j\omega_1)T(j2\omega_1)B(j2\omega_1)\cos 2\omega_1 t \\ + T^2(j\omega_2)T(j2\omega_2)B(j2\omega_2)\cos 2\omega_2 t \\ + T^2(j\omega_1)T(0)B(0) \\ + T^2(j\omega_2)T(0)B(0) \\ + 2T(j\omega_1)T(j\omega_2)T(j(\omega_1+\omega_2))B(j(\omega_1+\omega_2))\cos(\omega_1+\omega_2)t \\ + 2T(j\omega_1)T(j\omega_2)T(j(\omega_1-\omega_2))B(j(\omega_1-\omega_2))\cos(\omega_1-\omega_2)t \end{bmatrix}$$

$$\times \big[T(j\omega_1)\cos\omega_1 t + T(j\omega_2)\cos\omega_2 \big] \times T(\omega_{out}) \qquad (3.116)$$

The resulting 20 terms, each occurring at one of eight possible output frequencies are each multiplied by $T(j\omega)$ at the appropriate output frequency to account for the feedback through $B(s)$ and linear feedforward term k_1. The terms can be described by the following eight equations, each at a different output frequency.

$$v_{out31}(t)' = \frac{-v^3 k_2^2}{2} \left\{ \begin{array}{l} T^4(j\omega_1) T(j2\omega_1) B(j2\omega_1) \\[2mm] + 2T^4(j\omega_1) T(0) B(0) \\[2mm] + 2T^2(j\omega_1) T^2(j\omega_2) T(0) B(0) \\[2mm] + 2T^2(j\omega_1) T^2(j\omega_2) T(j(\omega_1+\omega_2)) B(j(\omega_1+\omega_2)) \\[2mm] + 2T^2(j\omega_1) T^2(j\omega_2) T(j(\omega_1-\omega_2)) B(j(\omega_1-\omega_2)) \end{array} \right\} \cos\omega_1 t \tag{3.117}$$

$$v_{out32}(t)' = \frac{-v^3 k_2^2}{2} \times \left\{ \begin{array}{l} + 2T^2(j\omega_1) T^2(j\omega_2) T(j(\omega_1+\omega_2)) B(j(\omega_1+\omega_2)) \\[2mm] + 2T^2(j\omega_1) T^2(j\omega_2) T(j(\omega_1-\omega_2)) B(j(\omega_1-\omega_2)) \\[2mm] + 2T^4(j\omega_2) T(j2\omega_2) B(j2\omega_2) \\[2mm] + 2T^4(j\omega_2) T(0) B(0) \\[2mm] + 2T^2(j\omega_1) T^2(j\omega_2) T(0) B(0) \end{array} \right\} \cos\omega_2 t \tag{3.118}$$

$$v_{out33}(t)' = \frac{-v^3 k_2^2}{2} \times \left\{ + T^3(j\omega_1) T(j2\omega_1) B(j2\omega_1) T(j3\omega_1) \right\} \cos 3\omega_1 t \tag{3.119}$$

$$v_{out34}(t)' = \frac{-v^3 k_2^2}{2} \times \left\{ + T^3(j\omega_2) T(j2\omega_2) B(j2\omega_2) T(j3\omega_2) \right\} \cos 3\omega_2 t \tag{3.120}$$

$$v_{out35}(t)' = \frac{-v^3 k_2^2}{2} \times \left\{ \begin{array}{l} + T(j\omega_1) T^2(j\omega_2) T(j2\omega_2) B(j2\omega_2) T(j(2\omega_2-\omega_1)) \\[2mm] + 2T(j\omega_1) T^2(j\omega_2) T(j(\omega_1-\omega_2)) B(j(\omega_1-\omega_2)) T(j(2\omega_2-\omega_1)) \end{array} \right\} \tag{3.121}$$
$$\times \cos(2\omega_2 - \omega_1)t$$

$$v_{out36}(t)' = \frac{-v^3 k_2^2}{2} \times \left\{ \begin{array}{l} + 2T^2(j\omega_1) T(j\omega_2) T(j(\omega_1-\omega_2)) B(j(\omega_1-\omega_2)) T(j(2\omega_1-\omega_2)) \\[2mm] + T^2(j\omega_1) T(j\omega_2) T(j2\omega_1) B(j2\omega_1) T(j(2\omega_1-\omega_2)) \end{array} \right\} \tag{3.122}$$
$$\times \cos(2\omega_1 - \omega_2)t$$

$$v_{out37}(t)' = \frac{-v^3 k_2^2}{2} \times \left\{ \begin{array}{l} + 2T(j\omega_1) T^2(j\omega_2) T(j2\omega_2) B(j(\omega_1+\omega_2)) T(j(2\omega_2+\omega_1)) \\[2mm] + T^2(j\omega_1) T(j\omega_2) T(j2\omega_1) B(j2\omega_1) T(j(2\omega_2+\omega_1)) \end{array} \right\} \tag{3.123}$$
$$\times \cos(2\omega_2 - \omega_1)t$$

$$v_{\text{out38}}(t)' = \frac{-v_3 k_2^2}{2} \times \left\{ \begin{array}{l} 2T^2(j\omega_1)T(j\omega_2)T(j(\omega_1+\omega_2))B(j(\omega_1+\omega_2))T(j(2\omega_1+\omega_2)) \\ +T^2(j\omega_1)T(j\omega_2)T(j2\omega_1)B(j2\omega_1)T(j(2\omega_1+\omega_2)) \end{array} \right\}$$

$$\times \cos(2\omega_1+\omega_2)t \qquad (3.124)$$

In step (5), v_{err1} is passed through the third-order nonlinearity resulting in

$$v_{\text{out3}}''(t) = \frac{k_3 v_1^3}{4} \left\{ \begin{array}{l} \left[3T^4(j\omega_1) + 6T^2(j\omega_2)T^2(j\omega_1) \right] \cos\omega_1 t \\[4pt] + \left[3T^4(j\omega_2) + 6T^2(j\omega_1)T^2(j\omega_2) \right] \cos\omega_2 t \\[4pt] + T^3(j\omega_1)T(j3\omega_1)\cos 3\omega_1 t \\[4pt] + T^3(j\omega_2)T(j3\omega_2)\cos 3\omega_2 t \\[4pt] + 3T^2(j\omega_1)T(j\omega_2)T(j(2\omega_1-\omega_2))\cos(2\omega_1-\omega_2)t \\[4pt] + 3T^2(j\omega_2)T(j\omega_1)T(j(2\omega_2-\omega_1))\cos(2\omega_2-\omega_1)t \\[4pt] + 3T^2(j\omega_1)T(j\omega_2)T(j(2\omega_1+\omega_2))\cos(2\omega_1+\omega_2)t \\[4pt] + 3T^2(j\omega_2)T(j\omega_1)T(j(2\omega_2+\omega_1))\cos(2\omega_2+\omega_1)t \end{array} \right\} \qquad (3.125)$$

Once more assume that

$$2\omega_1 \approx 2\omega_2 \approx \omega_2 + \omega_1 \approx 2\omega_0$$

$$\omega_0 = \frac{\omega_1 + \omega_2}{2} \approx \omega_1 \approx \omega_2 \approx 2\omega_2 - \omega_1 \approx 2\omega_1 - \omega_2$$

$$\omega_2 - \omega_1 = \Delta\omega \qquad (3.126)$$

$$3\omega_2 \approx 3\omega_1 \approx 2\omega_1 + \omega_2 \approx 2\omega_2 + \omega_1 \approx 3\omega_0$$

Using these narrowband approximations, the equations can be rewritten as

$$v_{\text{out1}}(t) = k_1 T(j\omega_0) v \cos\omega_1 t + k_1 T(j\omega_0) v \cos\omega_2 t \qquad (3.127)$$

$$v_{\text{out2}}(t) = \frac{v^2 k_2}{2} \left[\begin{array}{l} T^2(j\omega_0)T(j2\omega_0)\cos 2\omega_1 t \\[4pt] + T^2(j\omega_0)T(j2\omega_0)\cos 2\omega_2 t \\[4pt] + 2T^2(j\omega_0)T(0) \\[4pt] + 2T^2(j\omega_0)T(j2\omega_0)\cos(\omega_1+\omega_2)t \\[4pt] + 2T^2(j\omega_0)T(0)\cos(\omega_1-\omega_2)t \end{array} \right] \qquad (3.128)$$

$$v_{\text{out}31}(t)' = \frac{-v^3 k_2^2}{4} \left\{ \begin{array}{l} 3T^4(j\omega_0)T(j2\omega_0)B(j2\omega_0) \\ + 4T^4(j\omega_0)T(0)B(0) \\ + 2T^4(j\omega_0)T(j\Delta\omega)B(j\Delta\omega) \end{array} \right\} \cos\omega_1 t \qquad (3.129)$$

$$v_{\text{out}32}(t)' = \frac{-v^3 k_2^2}{4} \left\{ \begin{array}{l} 3T^4(j\omega_0)T(j2\omega_0)B(j2\omega_0) \\ + 2T^4(j\omega_0)T(j\Delta\omega)B(j\Delta\omega) \\ + 4T^4(j\omega_0)T(0)B(0) \end{array} \right\} \cos\omega_2 t \qquad (3.130)$$

$$v_{\text{out}33}(t)' = \frac{-v^3 k_2^2}{4} \left\{ T^3(j\omega_0)T(j2\omega_0)B(j2\omega_0)T(j3\omega_0) \right\} \cos 3\omega_1 t \qquad (3.131)$$

$$v_{\text{out}34}(t)' = \frac{-v^3 k_2^2}{4} \left\{ + T^3(j\omega_0)T(j2\omega_0)B(j2\omega_0)T(j3\omega_0) \right\} \cos 3\omega_2 t \qquad (3.132)$$

$$v_{\text{out}35}(t)' = \frac{-v^3 k_2^2}{4} \left\{ \begin{array}{l} + T^4(j\omega_0)T(j2\omega_0)B(j2\omega_0) \\ + 2T^4(j\omega_0)T(j\Delta\omega)B(j\Delta\omega) \end{array} \right\} \cos(2\omega_2 - \omega_1)t \qquad (3.133)$$

$$v_{\text{out}36}(t)' = \frac{-v^3 k_2^2}{4} \left\{ \begin{array}{l} + 2T^4(j\omega_0)T(j\Delta\omega)B(j\Delta\omega) \\ + T^4(j\omega_1)T(j2\omega_0)B(j2\omega_0) \end{array} \right\} \cos(2\omega_1 - \omega_2)t \qquad (3.134)$$

$$v_{\text{out}37}(t)' = \frac{-v^3 k_2^2}{4} \left\{ + 3T^3(j\omega_0)T(j2\omega_0)B(j2\omega_0)T(j3\omega_0) \right\} \cos(2\omega_2 + \omega_1)t \qquad (3.135)$$

$$v_{\text{out}38}(t)' = \frac{-v^3 k_2^2}{4} \left\{ + 3T^3(j\omega_0)T(j2\omega_0)B(j2\omega_0)T(j3\omega_0) \right\} \cos(2\omega_1 + \omega_2)t \qquad (3.136)$$

$$v_{\text{out}3}''(t) = \frac{k_3 v_1^3}{4} \left\{ \begin{array}{l} 9T^4(j\omega_0)\cos\omega_1 t \\ + 9T^4(j\omega_0)\cos\omega_2 t \\ + T^3(j\omega_0)T(j3\omega_0)\cos 3\omega_1 t \\ + T^3(j\omega_0)T(j3\omega_0)\cos 3\omega_2 t \\ + 3T^4(j\omega_0)\cos(2\omega_1 - \omega_2)t \\ + 3T^4(j\omega_0)\cos(2\omega_2 - \omega_1)t \\ + 3T^3(j\omega_0)T(j3\omega_0)\cos(2\omega_1 + \omega_2)t \\ + 3T^3(j\omega_0)T(j3\omega_0)\cos(2\omega_2 + \omega_1)t \end{array} \right\} \qquad (3.137)$$

The fundamental amplitude (assuming that it is weakly nonlinear) is simply

$$v_{o,fund} = k_1 T(j\omega_0) v \tag{3.138}$$

The in-band IM products at $(2\omega_2 - \omega_1)$ or $(2\omega_1 - \omega_2)$ will have an amplitude of

$$v_{0,IM} = \frac{3k_3 v_1^3}{4} T^4(j\omega_0) - \frac{v^3 k_2^2}{2} \left[T^4(j\omega_0) T(j2\omega_0) B(j2\omega_0) \right.$$
$$\left. + 2T^4(j\omega_0) T(j\Delta\omega) B(j\Delta\omega) \right] \tag{3.139}$$

This result can be tested with a resistor in feedback to compare to the previous example, and can be extended to an inductor in feedback. For a system with a simple feedback with $B(s) = R$, we have

$$v_{0,fund,R} = \frac{k_1 v}{1 + k_1 R} \tag{3.140}$$

and

$$v_{0,IM3,R} = \frac{3}{4} v^3 \left\{ \frac{k_3}{(1 + k_1 R)^4} - \frac{2R k_2^2}{(1 + k_1 R)^5} \right\} \tag{3.141}$$

If $B(s)$ is assumed to be an inductor:

$$B(s) = sL$$
$$T(s) = \frac{1}{1 + s k_1 L} \tag{3.142}$$

In this case:

$$v_{0,fund,L} = \frac{k_1 v}{1 + j\omega_0 k_1 L} \tag{3.143}$$

and

$$v_{0,IM3,L} = \frac{3k_3 v^3}{4} T^4(j\omega_0) - \frac{v^3 k_2^2}{2} \left[T^4(j\omega_0) T(j2\omega_0) B(j2\omega_0) + 2T^4(j\omega_0) T(j\Delta\omega) B(j\Delta\omega) \right]$$

$$= v^3 T^4(j\omega_0) \left[\frac{3k_3}{4} - \frac{k_2^2}{2} \left(T(j2\omega_0) B(j2\omega_0) + 2T(j\Delta\omega) B(j\Delta\omega) \right) \right] \tag{3.144}$$

$$= v^3 \left(\frac{1}{1 + j\omega_0 k_1 L} \right)^4 \left\{ \frac{3k_3}{4} - \frac{k_2^2}{2} \left(\frac{j2\omega_0 L}{1 + j2\omega_0 k_1 L} + 2\frac{j0.2\omega_0 L}{1 + j0.2\omega_0 k_1 L} \right) \right\}$$

Example 3.7:   Amplifier with Resistive and Inductive Emitter Degeneration

Use the theory developed in this section to do a numerical comparison of the third-order intermodulation terms for resistive and inductive degeneration for a common-emitter bipolar amplifier. The resistive degeneration case was previously shown in Example 3.6, and the schematic was shown in Figure 3.12. For this example, assume the parameters $R = 200$, $k_1 = 0.04$, $k_2 = 0.8$, $k_3 = 10.667$, $L = 34H$, $\omega_o = 2\pi * 1$ Hz, and $\Delta\omega = 2\pi * 0.06$ Hz.

Solution:

Note that at the fundamental frequency, $|Z_L|$ is 213.6Ω, slightly larger than the 200Ω impedance of the resistor. The following sample calculations are shown for $v_i = 0.05$V. For the simple feedback we have

$$v_{0,fund,R} = \frac{k_1 v}{1 + k_1 R} = \frac{0.04 \times 0.05}{1 + 0.04 \times 0.05} = \frac{2m}{1 + 8} = \frac{2m}{9} = 0.222 mV$$

$$v_{0,IM3,R} = \frac{3}{4} v^3 \left\{ \frac{k_3}{(1 + k_1 R)^4} - \frac{2Rk_2^2}{(1 + k_1 R)^5} \right\}$$

$$= \frac{3}{4} 0.05^3 \left\{ \frac{10.667}{9^4} - \frac{256}{9^5} \right\} = \frac{3}{4} 0.05^3 \{ 1.6258m - 4.335m \}$$

$$= 0.05^3 \cdot 2.032m = 254 nV$$

For inductive degeneration:

$$v_{0,fund,L} = \frac{k_1 v}{1 + j\omega_0 k_1 L} = \frac{0.04 * 0.05}{1 + j2\pi \cdot 0.04 \cdot 34} = \frac{2m}{1 + j8.5} = 0.2337 mV$$

$$v_{0,IM3,L} = v^3 \left(\frac{1}{1 + j\omega_0 k_1 L} \right)^4 \left\{ \frac{3k_3}{4} - \frac{k_2^2}{2} \left(\frac{j2\omega_0 L}{1 + j2\omega_0 k_1 L} + 2 \frac{j0.06\omega_0 L}{1 + j0.06\omega_0 k_1 L} \right) \right\}$$

$$= v^3 \left(\frac{1}{1 + j8.5} \right)^4 \left\{ 8 - 0.32 \left(\frac{j427.3}{1 + j17.1} + \frac{j25.64}{1 + j0.513} \right) \right\}$$

$$= v^3 \left(\frac{1}{1 + j8.5} \right)^4 \{ 8 - (11.3 + j6.96) \}$$

$$= v^3 (186.4\mu)(-3.3 - 6.96 j)$$

$$= 0.05^3 (186.4\mu)(7.7)$$

$$= 0.05^3 * 1.44m$$

$$= 180 nV$$

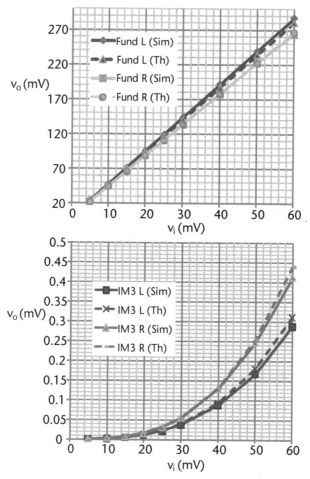

Figure 3.17 Fundamental and third-order intermodulation output voltages for resistive and inductive degeneration.

We note that the fundamental output voltage with the inductive degeneration is slightly larger than with equivalent resistive degeneration because of the phase shift of the inductor, that is, the magnitude of $1/(1 + X)$ is smaller than $1/(1 + jX)$. We also note that in spite of a larger fundamental voltage, the intermodulation voltage is lower for inductive degeneration. The fundamental and third-order output voltages are shown in Figure 3.17. These are compared to simulations showing good agreement even up to the maximum applied input of 60 mV. For larger input voltages, errors due to the approximations made in the derivations above would become more apparent.

3.5   Basic RF Building Blocks

In this book transistor level design will not be considered in any great detail, so in this book complete RF circuits will be the basic building blocks for all designs considered here. To that end, presented here is a list of basic RF building blocks and their properties that will need to be specified. It is a great asset to a systems designer to have some knowledge of the design of all these components as knowing the level of effort required to achieve a given performance target may lead to making a more optimal performance tradeoff.

3.5.1 Low Noise Amplifiers (LNAs)

LNAs are typically placed at the front of a radio to amplify weak signals while adding as little noise as possible. Typical specifications for an LNA are gain, noise figure, IP3, IP2, power consumption, input impedance, and 1-dB compression point. Typically, there is a compromise between performance parameters. As an example, the maximum possible frequency of operation is related first to the choice of technology. The transit frequency, f_T, can be used as an indicator of maximum operating frequency; typically, amplifiers might be operated up to one-tenth of the transit frequency. As typical numbers, a 32-nm CMOS processes may have a transit frequency f_T in the range of 200 GHz, while older technologies, such as 130-nm technology, f_T might be approaching 100 GHz, and 250-nm technology might have f_T of the order of 50 GHz. Bias current and hence power dissipation is also an important factor as maximum f_T is only achieved at relatively high current density. Another important factor in determining the maximum operating frequency is power dissipation. In general, operating at higher frequencies and achieving lower noise, higher linearity, and higher gain can be achieved at the expense of higher power dissipation. Thus, for any given technology, if high operating frequency is not critical, it may be possible to operate at a lower current density and hence a lower power dissipation. Because operating in the region of highest f_T is also close to the region for highest gain, best linearity, and lowest noise, it is apparent that power dissipation can be traded off for many performance parameters. As performance examples, in the literature, 60-GHz amplifier circuits have been designed in a 130-nm process, although being so close to f_T the gain was very low per stage, resulting in the need for multiple stages and relatively high bias current. More typically in a 130-nm process, center frequencies would be kept to less than 10 GHz, but with lower frequencies, there is the opportunity to operate at lower power dissipation.

A typical design might draw a few mA of bias current, have a noise figure of about 2 dB, and have a linearity (IIP3) of about –15 dBm. Linearity can be limited by the available swing at the output of the amplifier and this is made worse by having higher gain. Typical linearity, measured as IIP3, will range from about –20 dBm or even less for a low power designs to about 0 dBm with additional bias current and hence higher power dissipation.

3.5.2 Mixers

Mixers are frequency translation devices. They have two inputs and one output. For most operations in a radio, one of the inputs is driven with a sign wave or square wave reference signal often called a local oscillator (LO). The mixer multiplies this with a data input to translate the input data to the sum and difference of the reference and the input. The desired sideband or output frequency is then selected by filtering. Conversion gain in a mixer is defined as the amplitude of the output signal at the desired frequency divided by the amplitude of the input signal at the input frequency:

$$Conversion\ Gain = \frac{Sig_{out}(\omega_{desired})}{Sig_{in}(\omega_{in})} \qquad (3.145)$$

Most mixers operate by a switching mechanism that is equivalent to multiplying the input signal by plus or minus 1. This action has a maximum gain of $2/\pi$ or about −4 dB. This conversion gain is achieved only if the switches are large and are driven with a large enough local oscillator signal. Large switching components may result in larger parasitic capacitance, which may limit frequency response, or result in high levels of signal feedthrough. In a passive mixer, typically the input voltage is switched directly; thus, the switching conversion gain is also the mixer conversion gain. For an active mixer, for example, in a Gilbert cell mixer, the input voltage is converted to current through a transconductance stage and then the current is switched. Thus, in addition to switching conversion gain there is transconductance conversion gain that depends on the technology, transistor sizing, and bias levels. Generally, higher power dissipation is required to achieve higher gain.

Mixer noise figure is somewhat more complicated to define, compared to that of a simple gain stage, because of the frequency translation involved. Therefore, for mixers a slightly modified definition of noise figure is used. Noise factor for a mixer is defined as

$$F = \frac{N_{0tot}(\omega_{desired})}{N_{0(source)}(\omega_{desired})} \qquad (3.146)$$

where $N_{0tot}(\omega_{IF})$ is the total output noise power at the IF frequency and $N_{0(source)}$ $(\omega_{desired})$ is the output noise power at the output frequency due to the source. The source generates noise at all frequencies, and many of these frequencies will produce noise at the output frequency due to the mixing action of the circuit. Usually the two dominant frequencies are the input frequency and the image frequency (the frequency the same distance on the other side of the reference).

To make things even more complicated, *single-sideband* (SSB) noise figure, or *double-sideband* (DSB) noise figure are defined. The difference between the two definitions is the value of the denominator in (3.146). In the case of double-sideband noise figure, all the noise due to the source at the output frequency is considered (noise of the source at the input and image frequencies). In the case of single-sideband noise figure, only the noise at the output frequency, due to the source that originated at the RF frequency is considered. Thus, using the double-sideband noise figure definition, even an ideal noiseless mixer with a resistive termination at both the output and image frequencies would have a noise figure of 3 dB. This is because the noise of the source would be doubled in the output due to the mixing of both the RF and image frequency noise to the output frequency. Thus it can be seen that

$$N_{0(source)DSB} = N_{0(source)SSB} + 3\ dB \qquad (3.147)$$

and

$$NF_{DSB} = NF_{SSB} - 3\ dB \tag{3.148}$$

In practice this may not be quite correct, as an input filter will also affect the output noise, but this rule is usually used. Which definition is most appropriate depends on what type of radio architecture is employed and will be discussed in more detail later.

Largely because of the added complexity and the presence of noise that is frequency translated, mixers tend to be much noisier than LNAs. Because they are made from circuits just like amplifiers and commonly use active components they suffer from finite linearity just like an amplifier. In a passive mixer, the minimum possible noise is the loss through the system. Thus, the minimum double-sideband noise figure is about 4 dB since the loss in an ideal passive voltage-switched mixer is $2/\pi$ or about −4 dB. For active mixers there is further noise due to the gain stage, typically inversely proportional to the transconductance. Thus, as for gain, noise figure can typically be reduced by increasing the bias current and hence the power dissipation.

Typical specifications for a mixer are gain, noise figure (either SSB or DSB), IP3, IP2, power consumption, and 1-dB compression point. For a passive mixer, linearity is high if the switches are near perfect, and this is similar to the conditions for optimal gain and noise. For an active mixer, linearity can be limited by the linearity of an input transconductance stage, similar to the linearity conditions of an LNA. Also, similar to an LNA, if the mixer has high gain, clipping at the output may limit the linearity. Another consideration with mixers is how much isolation exists between ports. As with an LNA, there are trade-offs between linearity, noise, center frequency, bandwidth, gain, power dissipation, layout area, and cost.

3.5.3   Filters

Filters are frequency selective networks that pass certain frequencies and attenuate others. There are two major types that are usually used in radios: bandpass and lowpass. A bandpass filter passes a (usually narrow) band of frequencies and attempts to reject signals at all other frequencies. A lowpass filter passes frequencies up to some cutoff. Filters can be made of either active or passive components. Typical specifications are bandwidth of the passband, passband attenuation, and roll-off rate in the stopband. Filters have many practical limitations, for instance, LC filters are implemented with inductors that have finite quality factor that can be of the order of 20. Although positive feedback circuits can generate negative resistance to cancel the loss, resulting in very high Q filters at the expense of power dissipation, these circuits will then suffer from finite linearity, potential instability, and noise compared to passive LC circuits. A different technique to achieving high Q is to use sampling techniques. At low frequencies, it is straightforward to achieve a narrow bandwidth, but then this narrow bandwidth is translated to higher frequencies through the sampling operation resulting in a high Q filter.

3.5.4 Voltage-Controlled Oscillators and Frequency Synthesizers

A voltage-controlled oscillator (VCO) is a signal source whose frequency can be controlled by a dc voltage. VCOs or more commonly VCOs embedded in a frequency synthesizer will often provide any needed local oscillator (LO) signals in a radio. A complete frequency synthesizer (which will be the subject of Chapter 6) will often in addition to the VCO's properties specify maximum sideband spur levels, tuning range, tuning resolution or step size, integrated phase noise.

There are many different types of oscillators. LC oscillators are typically selected for low phase noise (and correspondingly low jitter) as well as low power dissipation, but they suffer from a relatively narrow tuning range determined by the range of variable capacitors (e.g., varactors) or switched components. Another disadvantage of LC oscillators is that they can be quite large when implanted in an IC technology because of the inductors which are typically a few hundred microns in diameter for inductors in the nanohenry range. Ring oscillators can have broad tuning range and much smaller layout areas compared to LC oscillators, but typically have more noise. Two or four stage ring oscillators are often chosen since they can provide outputs with 90° of phase shift between them. Distributed oscillators, rotary traveling-wave oscillators, and standing wave oscillators make use of transmission lines or artificial transmission lines to achieve very low noise, similar to an LC oscillator. The power consumption per stage is of the same order as an LC oscillator. The advantage over an LC oscillator is that it is possible to achieve multiple output phases. The disadvantage is that of large layout area due to the transmission lines. As well, achieving a wide tuning range is also challenging.

Noise in oscillators goes up roughly in proportion to the carrier frequency and inversely proportional to the offset frequency from the carrier. Generally, higher power dissipation can be used to achieve higher output signal levels and lower phase noise. Higher tuning range in a single band typically results in higher noise, so, for a wide tuning range it is important to have this in multiple bands.

3.5.5 Variable Gain Amplifiers

Some amplifiers need to have gain levels that are programmable. They have all the properties of any standard amplifier, but in addition have gain levels and gain steps (the distance between two gain levels) must be specified. Typically the requirement for adjustable gain results in a compromise with other performance parameters, for example, noise or linearity. For example, the noise will likely be higher and thus the first stage of an LNA would usually not be designed to have variable gain. Similarly, linearity may suffer and so variable gain is typically not used for the final stage of a power amplifier.

3.5.6 Power Amplifiers

Power amplifiers (PAs) are responsible for delivering a required amount of power to the antenna. Power amplifiers usually consume a large amount of DC power so the efficiency at which they operate is often very important. Efficiency η, sometimes

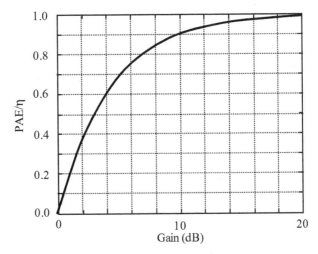

Figure 3.18   Normalized power-added efficiency versus gain.

also called *dc-to-RF efficiency*, is the measure of how effectively power from the supply is converted into output power and is given by

$$\eta = \frac{P_{\text{out}}}{P_{\text{dc}}} \tag{3.149}$$

where P_{out} is the ac output power and P_{dc} is the power consumed from the supply. *Power-added efficiency* (PAE) is similar to efficiency; however, it takes the gain of the amplifier into account as follows:

$$\text{PAE} = \frac{P_{\text{out}} - P_{\text{in}}}{P_{\text{dc}}} = \frac{P_{\text{out}} - P_{\text{out}}/G}{P_{\text{dc}}} = \eta\left(1 - \frac{1}{G}\right) \tag{3.150}$$

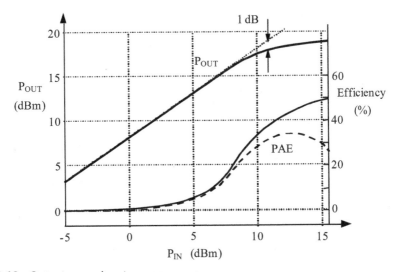

Figure 3.19   Output versus input power.

where G is the power gain defined as P_{out}/P_{in}. Thus, it can be seen that for high gain, power-added efficiency PAE is the same as dc-to-RF efficiency η. Figure 3.18 shows the efficiency in comparison to power-added efficiency for a range of power gains. It can be seen that for gain higher than 10 dB, PAE is within 10% of the efficiency η. As the gain compresses, PAE decreases. For example, if the gain is 3 dB, the PAE is only half of the dc-to-RF efficiency.

A typical plot of output power and efficiency versus input power is shown in Figure 3.19. It can be seen that while efficiency keeps increasing for higher input power, as the amplifier compresses and gain decreases, the power-added efficiency also decreases. Thus, there is an optimal value of power-added efficiency and it typically occurs a few decibels beyond the 1-dB compression point.

When specifying a PA, often the most important parameters are efficiency, PAE, output power, output 1-dB compression point, OIP3, gain, and output impedance. Efficiency is typically traded off with linearity. That is, as shown in Figure 3.19, maximum efficiency is achieved well beyond the linear region of operation. In an earlier section, nonlinearity was described with a power series. This may be augmented with Volterra series analysis in the likely case that there are memory effects, for example due to the presence of reactive components such as capacitors and inductors. It should be noted that such analysis is valid only for weak nonlinearity and so is not completely valid for power amplifiers, but it is often used as a starting point and detailed simulations are done to refine the design. In the trade-off between linearity and efficiency, different classes of amplifiers are sometimes chosen according to how the amplifiers are biased. For example, class A amplifiers are biased to have current flowing though the driver transistors at all times, so are inherently the most linear, but the least efficient. In contrast, classes AB, B, or C have current flowing for only part of the time and hence are less linear but are more efficient. Modern amplifiers are often designed using linearity enhancement schemes such as feedforward, feedback, or predistortion techniques, or as a combination of amplifiers that provide improved performance compared to a single amplifier. An example of this is the Doherty amplifier that has a main amplifier operating in class AB or class B, and an auxiliary amplifier that operates in class C, but only when the signal is large. Through this combination, efficiency is enhanced over a broader range of input power compared to the individual amplifiers.

3.5.7 Phase Shifters

Often phase shifters are used to implement a phase shift in an LO tone. A common requirement is to have two copies of the input signal at 90° relative phase shift. Such pairs of phase-shifted tones are commonly labeled I and Q for in phase and quadrature phase. Common specifications for a phase shifter are power consumption, phase shift, and expected phase error or mismatch. Phase shifters may be designed based on circuit components such as resistors, capacitors and inductors, as delay lines or implemented as a divide-by-two circuit. In any case, additional buffers may be required to compensate for signal attenuation; thus, additional power dissipation is expected. There may also be errors due to process variations. All of these devices also have a limited bandwidth of operation. In a filtering approach, one way to reduce the error and to increase the bandwidth is to use more stages, that is, to use cascaded filter sections.

Table 3.6 Common RF Building Blocks

Name	Symbol	Common Important Specifications
Low noise amplifier (LNA)	LNA RF_{in} —▷— RF_{out}	Power consumption, input impedance, IP3, IP2, 1-dB compression point, gain, NF, reverse isolation
Mixer	Mixer Sig_{in} —⊗— Sig_{out} LO_{in}	Power consumption, IP3, IP2, 1dB compression point, NF (either DSB or SSB), LO_{in} to Sig_{in} isolation, LO_{in} to Sig_{out} isolation
Filter	Filter Sig_{in} —▨— Sig_{out}	Bandwidth, stopband roll off rate, linearity (if it is active)
Voltage-controlled oscillator (VCO) or frequency synthezier	LO_{out}	Power consumption, gain, operating frequency range, phase noise, spur levels, frequency step size, integrated phase noise
Variable gain amplifier (VGA)	VGA Sig_{in} —▷— Sig_{out}	Power consumption, IP3, IP2, 1dB compression point, gain, gain steps, bandwidth, noise
Power amplifier (PA)	PA RF_{in} —▷— RF_{out}	Efficiency, PAE, output power, output 1-dB compression point, OIP3, gain, output impedance
LO phase shifter	I LO_{in}—[90°] Q	Phase shift, power consumption, IQ mismatch
Digital-to-analog (D/A) and analog-to-digital (A/D) converters	A/D An_{in}—[⎍]—Dig_{out} D/A Dig_{in}—[⎍]—An_{out}	Number of bits, sampling frequency, clock jitter, power consumption dynamic range, maximum input/output signal level
RF receive/transmit switch	Rx —Switch Tx	Isolation, linearity, loss
Antenna	▽	Gain, impedance, bandwidth of operation, directivity

As each stage has associated signal loss, more amplification will be required, resulting in an increase in power dissipation and with possible noise implications.

3.5.8 Analog-to-Digital (A/D) and Digital-to-Analog (D/A) Converters

A/D and D/A converters are used as the interface between the analog radio and the digital base band. Depending on the nature of a project they may be implemented with the radio or with the DSP core. Generally, the specifications for these parts are power consumption, number of bits, sampling frequency, and the clock jitter of the reference used to drive them. Operating them at a higher speed usually requires higher power dissipation and typically results in an effective lower number of bits. At the highest speed end of A/D converters are flash converters that can operate at multiple gigasamples per second, but use large layout area and higher power dissipation, and typically have only a few bits. At the low-frequency end are oversampling A/D converters, which can have a high number of bits, for example, over 20, and typically have sampling frequencies in the mega-sample per second range. Oversampled converters can have baseband or bandpass sampling. Other types of A/D converters are successive approximation and pipelined converters or combination types. When specifying D/A converters, linearity power dissipation and maximum signal size may be additional concerns.

3.5.9 RF Switch

A switch is often needed to select between a transmit path and a receive path in a radio. Such a switch will need to be low loss and have good isolation between the three terminals. If the switch is made from active components, linearity may be an issue as well.

3.5.10 Antenna

The antenna is what changes the RF energy into EM radiation that can be transmitted through the air and vice versa. Antennas have the ability to trade off directivity (how much energy is transmitted in a given direction) with gain. Thus, a highly directive antenna can have a lot more gain than one that transmits in almost every direction.

Table 3.6 summarizes the discussion in this section and lists most common RF building blocks and some of their most important properties.

References

[1] Papoulis, A., *Probability, Random Variables, and Stochastic Processes*, New York: McGraw-Hill, 1984.

[2] Gray, P. R., et al., *Analysis and Design of Analog Integrated Circuits*, 4th ed., New York: John Wiley & Sons, 2001.

[3] Stremler, F. G., *Introduction to Communication Systems,* 3rd ed., Reading, MA: Addison-Wesley Publishing, 1990.

[4] Jordan, E. C., and K. G. Balmain, *Electromagnetic Waves and Radiating Systems*, 2nd ed., Englewood Cliffs, NJ: Prentice-Hall, 1968.

[5] Rappaport, T. S., *Wireless Communications*, Upper Saddle River, NJ: Prentice-Hall, 1996.

[6] Proakis, J. G., *Digital Communications*, 3rd ed., New York: McGraw-Hill, 1995.

[7] Gonzalez, G., *Microwave Transistor Amplifiers*, 2nd ed., Upper Saddle River, NJ: Prentice-Hall, 1997.

[8] Wambacq, P., and W. Sansen, *Distortion Analysis of Analog Integrated Circuits*, Boston, MA: Kluwer Academic Publishers, 1998.

[9] Wambacq, P., et al., "High-Frequency Distortion Analysis of Analog Integrated Circuits," *IEEE Transactions on Circuits and Systems II: Analog and Digital Signal Processing*, Vol. 46, No. 3, March 1999, pp. 335–345.

[10] "Some Notes on Composite Second and Third Order Intermodulation Distortions," *Matrix Technical Notes MTN-108,* December 15, 1998.

[11] "The Relationship of Intercept Points and Composite Distortions," *Matrix Technical Notes MTN-109,* February 18, 1998.

CHAPTER 4

System-Level Architecture

4.1 Introduction

In this chapter, various configurations of circuit building blocks used to form radios are discussed. Some of the following sections deal with block diagrams for complete transceivers, while other sections are about subsystems that solve particular problems. Although it would be impossible to discuss all architectures here, common ones are presented and such discussion includes many of the basic concepts common to most architectures.

4.2 Superheterodyne Transceivers

A block diagram of a typical half duplex superheterodyne radio transceiver is shown in Figure 4.1 [1–7]. Modulated signals are transmitted and received at some frequency by an antenna. If the radio is receiving information, then the signals are passed from the antenna to the receiver (Rx) part of the radio. If the radio is transmitting, then signals are passed to the antenna from the transmitter (Tx) part of the radio. Radios either transmit and receive at the same time (called a *full-duplex transceiver*) or alternate between transmitting and receiving (called a *half-duplex transceiver*). In a half-duplex transceiver, it is possible to put a switch between the antenna and the Rx and Tx paths to provide improved isolation, while in a full-duplex transceiver, the switch must be omitted and the two input filters (usually called a *duplexor*) have the sole responsibility of isolating the Tx and Rx paths without the aid of a switch.

In the Rx path, the signals are first passed through a preselect bandpass filter to remove interference from frequency bands other than the one of interest. Although the preselect BPF will negatively impact the noise figure of the radio, placing a filter here makes the design of the receiver input easier as it will reduce the power levels with which the front end must deal. The signal is then amplified by a *low noise amplifier* (LNA) to increase the power in weak signals, while adding as little noise (unwanted random signals) as possible. This amplifier may include some form of adjustable gain or gain steps. Adding a gain step here complicates the design of the LNA, but allows the gain to be reduced for larger input levels, thus reducing the linearity requirements of the rest of the radio. The spectrum is then further filtered by an image filter. This is required because the preselect BPF will typically not provide enough image rejection, or may have been omitted for noise reasons. The signal is then downconverted by a mixer to an *intermediate frequency* (IF). The IF frequency must be chosen with great care, taking into account many factors including

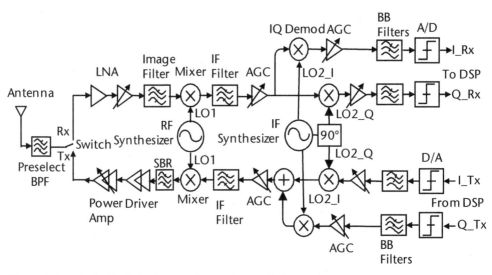

Figure 4.1 A typical half-duplex superheterodyne radio transceiver.

interaction of spurs and mixing of LO harmonics. Generally, a higher IF frequency will make the job of the image filter easier, while making the design of the IF stage and especially the IF filter harder as it will need to work at a higher frequency.

The mixer (also sometimes called a multiplier) mixes the incoming *radio-frequency* (RF) signal with the output from the RF frequency synthesizer which is an accurate frequency reference generator also called a *local oscillator* (LO). The LO is tuned so that the frequency of the desired IF signal is always at the same frequency. The LO can be either low-side injected (the LO is at a frequency less than the RF frequency) or high-side injected (the LO is at a frequency greater than the RF frequency). For low-side injection, the IF frequency is given by

$$f_{IF} = f_{RF} - f_{LO} \qquad (4.1)$$

while for high-side injection it is

$$f_{IF} = f_{LO} - f_{RF} \qquad (4.2)$$

As the IF stage is usually at a fixed frequency, the synthesizer must be programmable so that it can be tuned to whatever input frequency is desired at a given time. An input signal at an equal distance from the LO on the other side from the desired RF signal is called the *image signal*. It is called the image because a signal at this frequency after mixing will be at the same IF as the desired signal. Therefore, the image signal cannot be removed by filtering after mixing has taken place. Thus, an important job of the RF filters is to remove any image signals before such mixing action takes place. This is illustrated in Figure 4.2. It is also important to keep any LO spurs as low as possible. For example, in Figure 4.2, the LO spurs illustrated will act to convert channels 2 and 6 into the IF frequency band. If these spurs are too strong, then this can corrupt the downconversion process as well. Also, after mixing, there is no further filtering that can fix this problem.

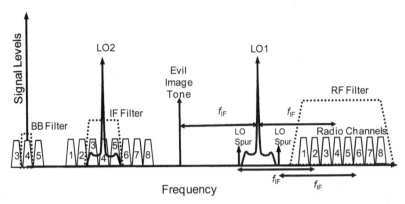

Figure 4.2   Figure showing radio receiver frequency plan. As shown, the radio is tuned to receive channel 4 and has a low-side injected LO.

After mixing to an IF, additional filtering is usually performed. At the IF, unwanted channels can be filtered out as now the desired channel is always centered at the same known frequency, leaving only the channel of interest now centered at the IF, and possibly some subset of the original adjacent channels depending on the quality of the filter used. Usually, *automatic gain control* (AGC) amplifiers are also included at the IF. They adjust the gain of the radio so that its output amplitude is always constant. AGC is performed at the IF to avoid mismatches in gain between the I and Q paths once the signal reaches baseband. Once through the AGC, the signals are downconverted a second time to baseband (the signals are now centered around DC or zero frequency). This second downconversion requires a second frequency synthesizer that produces both 0° and 90° output signals at the IF frequency. Two mixers downconvert the signals into *in-phase* (I) and *quadrature phase* (Q) paths. By using two separate paths, both amplitude and phase information can be recovered, and as a result, the incoming phase of the RF signal does not need to be synchronized to the phase of the LO tone. The I and Q signals are then passed through baseband filters which remove the rest of the unwanted channels. Finally, the signal is passed through an analog-to-digital converter and into the back end of the radio. There may be additional, possibly programmable, gain stages in the baseband. The better the quality of baseband filter the easier the job of the ADC as a good quality BB filter will remove more unwanted out-of-band energy than a poor quality one. Thus, there is a direct tradeoff between the performance levels of these two components. Further signal processing is performed in the *baseband signal processing* (BBSP) circuitry in the back end of the radio.

The transmitter works much the same way except in reverse. The BBSP circuitry followed by the digital-to-analog converters (DAC) produces signals in quadrature. These signals are then filtered (presumably the spectrum of the signals has already been shaped in the BBSP section of the radio, but harmonics or other unwanted tones can be generated by the DAC) and upconverted to an IF frequency where they are summed into one signal. The Tx will usually have some AGC function which may be either in the baseband or the IF stage. Placing AGC in the baseband creates matching issues, while implementing it at higher frequencies will make the circuit implementation harder. The IF signal is upconverted to the RF frequency by the mixer. If the LO is low-side injected the mixer is used to generate sum, rather than difference products. Thus, for low-side injection, the RF frequency is given by

$$f_{RF} = f_{LO} + f_{IF} \qquad (4.3)$$

and if the LO is high-side injected, the frequency of the RF signal is given by

$$f_{RF} = f_{LO} - f_{IF} \qquad (4.4)$$

Note that the mixer in either case will create an unwanted sideband as both the sum and difference frequencies will be generated, regardless of which one is actually wanted. Once upconverted to RF, the signal is passed through a sideband select filter (SBS) to remove the unwanted sideband and any LO feed through as well. Ideally, this is done before the PA to avoid using power to amplify an unwanted signal. A power amplifier is next used to increase the power of the signals. After the PA, additional filters may be present to remove PA harmonics and make sure that transmit power masks are not violated (transmit power masks specify the maximum allowed radiated power outside the desired channel bandwidth). The RF signal is then radiated by the antenna into the air. In the RF section, the PA itself may have a power control function or additional AGC. If the power level is constant, it must be high enough so that the signal can be detected at the maximum distance dictated by the system specifications.

Another refinement that can be made to the superheterodyne architecture is to design the receiver to have two IFs rather than just one as shown in Figure 4.3. The main advantage to this refinement to the architecture is seen in the receiver. In this case, the first IF can be chosen to be at a higher frequency easing the requirements of the image reject filter at RF, while the second IF can be chosen to be at a much lower frequency, allowing the second IF filter to achieve better channel selection with a lower-quality filter. Additionally, the final downconversion and IQ generation can be performed at a lower frequency, making this task easier. The disadvantage to this architecture is that it requires more circuitry and the addition of yet another synthesizer making this a much less popular choice at the time of this writing.

In a full-duplex radio, the Tx and Rx operate at different frequencies; therefore, it is no longer possible to share all the synthesizers. Usually, it is the RF synthesizer that is shared and the two IF synthesizers which are separated. A full-duplex radio

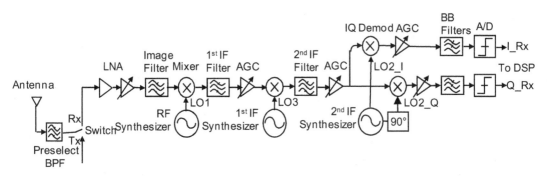

Figure 4.3 A half-duplex superheterodyne radio receiver with two IF stages.

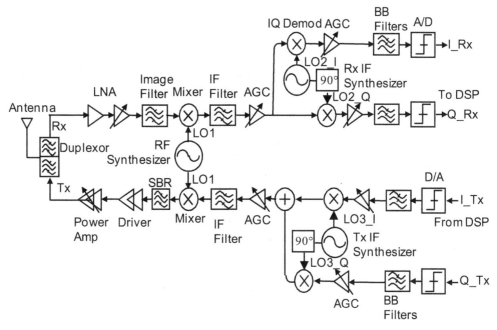

Figure 4.4   A typical full-duplex superheterodyne radio transceiver.

diagram is shown in Figure 4.4. Note that if further isolation were needed in a full-duplex application, then one might employ two antennas, but because these antennas would likely have to be close to each other, this would not remove the need for a duplexor as energy radiated by the transmitting antenna would be absorbed the receiving antenna.

4.3   Direct Conversion Transceivers

A half-duplex direct downconversion radio architecture is shown in Figure 4.5. In this architecture, the IF stage is omitted and the signals are converted directly to DC. For this reason, the architecture is sometimes called a *zero-IF radio*. The direct-conversion transceiver has become popular in recent years because it saves the area and power associated with a second synthesizer, although if it was made into a full-duplex radio, two RF synthesizers would still be needed. It also requires fewer filters making it more compatible with modern IC technologies. In a direct conversion transceiver, no image filter is required and the LO frequency selection becomes trivial. However, generating I and Q signals from a synthesizer at higher frequencies is much more difficult than doing so at the IF. Because the LO signal is now at the same frequency as the incoming RF signal, LO energy can couple into the RF path and cause problems. Without an IF stage more gain and gain control must be done at baseband, making the amplitude and phase matching of both the I and Q paths more difficult than in the case of the superheterodyne radio. The process of downconversion is illustrated in Figure 4.6. Once again, LO spurs can be troublesome in a direct downconversion radio. Here

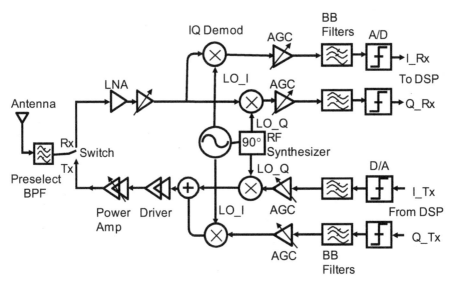

Figure 4.5 A half-duplex direct downconversion radio.

the two sideband spurs will act to convert unwanted channels 2 and 6 to baseband, possibly corrupting the desired signal.

The transmit path is also simpler than the superheterodyne case. Here the signal is converted directly to RF. This may mean that part or all of the AGC function will need to be done at RF to avoid matching problems. However, in this case, the RF filter requirements on the transmitter are greatly reduced as direct conversion means that there is no unwanted sideband to be filtered out or LO feedthrough to worry about. Thus, in some cases, it may be advantageous to combine a superheterodyne receiver with a direct conversion transmitter. Another problem with direct conversion architecture when used in a full-duplex radio is that there would be two high-frequency VCOs operating at close to the same frequency. These VCOs would likely try to pull each other off frequency, which would be a difficult problem to overcome. One VCO would likely have to operate at another frequency band and a frequency translation network would be needed to generate the final LO frequency.

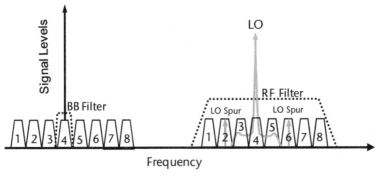

Figure 4.6 The radio receiver frequency plan. As shown, the radio is tuned to receive channel 4.

4.4   Offset Phase Locked Loop (PLL) Transmitters

Another variant on transmitter architecture is the offset phase lock loop, as shown in Figure 4.7. This architecture works very well with constant envelope modulations and has numerous advantages. The transmit path starts in the traditional way, but once the signal is at IF, all the amplitude information is stripped away via use of a limiter and the signal is fed into a phase detector. The phase detector, mixer, LPF, and RF VCO form a feedback loop that forces the phase of the VCO output to track the phase of the data. This feedback loop modulates the VCO. Note that the mixer is included in the feedback path so that the IF frequency is equal to the difference in the frequency of the transmitter RF synthesizer and the RF VCO. The mixer is used instead of a divider in this architecture to convert the VCO frequency back down because using a divider will not divide the phase of the VCO output; thus, the phase of the data can be upconverted without being multiplied by the difference in frequency between the data and the RF VCO. The LPF is included to adjust the feedback system parameters. The RF VCO can then directly drive the PA. In this case, because phase-only modulation is used, the PA can be implemented as a highly efficient nonlinear device. Note that PLLs will be studied in much more detail later.

The OPLL has a number of advantages. First, because the VCO drives the PA directly, the out-of-channel noise floor can be much lower than when a mixer drives the PA. This can be a very big deal in a full-duplex radio in which energy leakage into the Rx band can be problematic. As well, removing the mixer from the Tx path means that there is no other sideband to be filtered. Thus, the loss associated with the filter and the bother of implementing it are removed. (RF filters are much harder to implement than baseband filters.)

Engineers were so impressed with the performance advantages of this architecture that they began to wonder if it could be used with more advanced modulation types. This led to the idea of adding AM modulation back into the waveform by

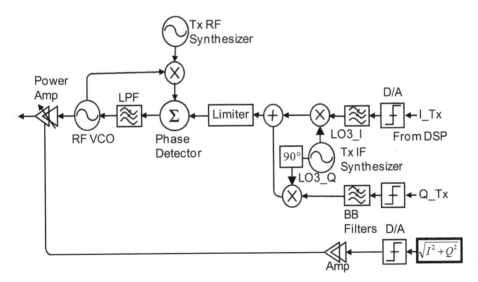

Figure 4.7   OPLL Transmitter Architecture.

dynamically adjusting the power supply of the PA. This is done from the base-band by taking the modulation amplitude, passing it through a digital-to-analog converter, amplifying it, and applying the signal to the PA. In addition to all the advantages previously discussed, it is now also possible to have AM modulation on top of a phase modulated signal with a highly efficient nonlinear PA.

4.5 Low IF Transceiver

Another alternative architecture is called a low IF transceiver [5]. This is basically a superheterodyne radio, but in this design the IF frequency is chosen to be so low that both the IF and baseband processing can be performed in the digital domain. To make the analog-to-digital converter implementation easier, the IF should be chosen as low as possible and can be as low as the signal bandwidth. Contrary to the offset PLL, the low IF architecture sees most of its advantages on the receive side. The RF circuitry can be as simple as that of a direct downconversion radio, but this radio will not suffer from the DC offset problem and flicker noise. The main problem with a low IF receiver is that the image frequency will now be very close to the desired sideband, making it almost impossible to filter out. Thus, an image-reject architecture needs to be employed. An example of a low IF receiver is shown in Figure 4.8.

This receiver solves the image problem by implementing a Weaver image reject architecture (discussed in more detail in Section 4.10). The advantage here is that the second set of mixers and LO signal is implemented in the BBSP. Thus, the phase shift of the second set of LOs will be perfect. As the analog parts will suffer from imperfections, this can also be calibrated out in the BBSP. Two amplifiers with gain α and β are implemented to fix both amplitude and phase mismatch problems introduced by the analog hardware. If there is no mismatch at all, then $\alpha = 1$ and $\beta = 0$. To calibrate the receiver, first it is placed in a test mode where a signal from the transmitter is fed back to the receiver. The first test is performed with $\alpha = 1$ and $\beta = 0$. In this test, by adjusting the sign of the last digital addition, both the desired

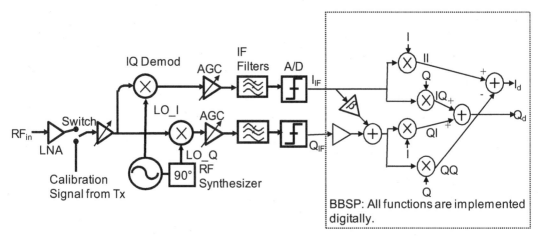

Figure 4.8   A Low IF receiver configured for low-side LO injection.

and image sidebands can be recovered. Assume that the RF paths have some amplitude mismatch ΔA and that there is some phase error in the RF synthesizer ϕ_e. Now because correction is applied to the Q path only, assume that these errors occur in the Q path and that the I path is ideal. Thus, assuming that the RF is high-side injected,

$$I_{IF} = \cos\left(\omega_{IF}t + \varphi(t)\right)$$

$$Q_{IF} = -\left(1 + \Delta A\right)\sin\left(\omega_{IF}t + \phi_e + \varphi(t)\right)$$

(4.5)

Note that because these are now digital signals, they can be normalized to an amplitude of unity. The result after the second downconversion is

$$II = \frac{1}{2}\cos\varphi(t)$$

$$IQ = -\frac{1}{2}\sin\varphi(t)$$

$$QI = -\frac{1+\Delta A}{2}\sin\left[\varphi(t) + \phi_e\right]$$

$$QQ = -\frac{1+\Delta A}{2}\cos\left[\varphi(t) + \phi_e\right]$$

(4.6)

Now we can determine the differential signals:

$$I_d = II - QQ = \frac{1}{2}\cos\varphi(t) + \frac{1}{2}\cos\left[\varphi(t) + \phi_e\right] + \frac{\Delta A}{2}\cos\left[\varphi(t) + \phi_e\right]$$

$$Q_d = IQ - QI = \frac{1}{2}\sin\varphi(t) + \frac{1}{2}\sin\left[\varphi(t) + \phi_e\right] + \frac{\Delta A}{2}\sin\left[\varphi(t) + \phi_e\right]$$

(4.7)

The last term in the expression will be small assuming that $\Delta A \ll 1$, so

$$I_d \approx \frac{1}{2}\cos\varphi(t) + \frac{1}{2}\cos\left[\varphi(t) + \phi_e\right]$$

$$Q_d \approx \frac{1}{2}\sin\varphi(t) + \frac{1}{2}\sin\left[\varphi(t) + \phi_e\right]$$

(4.8)

Making use of the following trig identities:

$$\cos a + \cos b = 2\cos\left(\frac{a+b}{2}\right)\cdot\cos\left(\frac{a-b}{2}\right)$$

$$\sin a + \sin b = 2\sin\left(\frac{a+b}{2}\right)\cdot\cos\left(\frac{a-b}{2}\right)$$

(4.9)

the previous equations can be rewritten as

$$I_d = II - QQ \approx \cos\left(\frac{2\varphi(t) + \phi_e}{2}\right)\cos\left(\frac{\phi_e}{2}\right) \approx \cos\varphi(t)\cos\left(\frac{\phi_e}{2}\right)$$

$$Q_d = IQ + QI \approx \sin\left(\frac{2\varphi(t) + \phi_e}{2}\right)\cos\left(\frac{\phi_e}{2}\right) \approx \sin\varphi(t)\cos\left(\frac{\phi_e}{2}\right)$$

(4.10)

If we reverse the signs of the final addition, we can set the mixer to recover the other sideband. If we are still feeding in the same signal at RF, it will now be at the image frequency for this new BBSP configuration. Thus, we can evaluate how much image gets through the mixers. The image signal can also be determined (remembering to reverse the additions and subtractions so now the mixer is trying to reject the RF signal as it is the unwanted sideband) as

$$I_{im} = II + QQ = \frac{1}{2}\cos\varphi(t) - \frac{1}{2}\cos[\varphi(t) + \phi_e] - \frac{\Delta A}{2}\cos[\varphi(t) + \phi_e]$$

$$Q_{im} = IQ - QI = -\frac{1}{2}\sin\varphi(t) + \frac{1}{2}\sin[\varphi(t) + \phi_e] + \frac{\Delta A}{2}\sin[\varphi(t) + \phi_e]$$

(4.11)

Making use of the following trig identities:

$$\cos(a - b) = \cos(a)\cdot\cos(b) + \sin(a)\cdot\sin(b)$$

$$\cos(a) - \cos(b) = -2\sin\left(\frac{a+b}{2}\right)\cdot\sin\left(\frac{a-b}{2}\right)$$

$$\sin(a) - \sin(b) = 2\cos\left(\frac{a+b}{2}\right)\cdot\sin\left(\frac{a-b}{2}\right)$$

$$\sin(a + b) = \cos(a)\cdot\sin(b) + \sin(a)\cdot\cos(b)$$

(4.12)

the previous expressions can be manipulated to give

$$I_{im} = \frac{1}{2}\left[-2\sin\left(\frac{\varphi(t) + \varphi(t) + \phi_e}{2}\right)\sin\left(\frac{\varphi(t) + \phi_e - \varphi(t)}{2}\right) - \Delta A\cos(\phi_e)\cos\varphi(t)\right.$$

$$\left. - \Delta A\sin(\phi_e)\sin\varphi(t)\right]$$

$$Q_{im} = \frac{1}{2}\left[2\cos\left(\frac{\varphi(t) + \varphi(t) + \phi_e}{2}\right)\sin\left(\frac{\varphi(t) - \varphi(t) - \phi_e}{2}\right) + \Delta A\cos(\phi_e)\sin\varphi(t)\right.$$

$$\left. + \Delta A\sin(\phi_e)\cos\varphi(t)\right]$$

(4.13)

Simplifying results in

$$I_{im} \approx -\frac{1}{2}\left[-\Delta A \cos(\phi_e)\cos\varphi(t) - 2\sin\left(\frac{\phi_e}{2}\right)\sin\varphi(t)\right]$$

$$Q_{im} \approx \frac{1}{2}\left[\Delta A \cos(\phi_e)\sin\varphi(t) - 2\sin\left(\frac{\phi_e}{2}\right)\cos\varphi(t)\right] \tag{4.14}$$

Now the ratio of the image to the calibration tone can be expressed as

$$\frac{I_{im}}{I_d} = -\frac{1}{2}\left[-\Delta A \frac{\cos(\phi_e)}{\cos\left(\frac{\phi_e}{2}\right)} - 2\tan\left(\frac{\phi_e}{2}\right)\tan\varphi(+)\right] \approx -\frac{1}{2}\left[-\Delta A - 2\tan\left(\frac{\phi_e}{2}\right)\tan\varphi(+)\right]$$

$$= \frac{1}{2}\left[\Delta A + 2\tan\left(\frac{\phi_e}{2}\right)\frac{Q_d}{I_d}\right]$$

$$\frac{Q_{im}}{Q_d} = \frac{1}{2}\left[\Delta A \frac{\cos(\phi_e)}{\cos\left(\frac{\phi_e}{2}\right)} - 2\tan\left(\frac{\phi_e}{2}\right)\cot\varphi(t)\right] \approx \frac{1}{2}\left[\Delta A - 2\tan\left(\frac{\phi_e}{2}\right)\cot\varphi(t)\right]$$

$$= \frac{1}{2}\left[\Delta A - 2\tan\left(\frac{\phi_e}{2}\right)\frac{I_d}{Q_d}\right] \tag{4.15}$$

where

$$\frac{I_d}{Q_d} = \frac{\cos(\varphi(t))}{\sin(\varphi(t))} \tag{4.16}$$

Solving (4.15) for the errors will allow us to determine their values as

$$\Delta A = \frac{-2(Q_d Q_{im} + I_d I_{im})}{I_d^2 + Q_d^2} \tag{4.17}$$

and

$$\phi_e = 2\tan^{-1}\left[\frac{Q_d I_{im} - I_d Q_{im}}{Q_d^2 + I_d^2}\right] \tag{4.18}$$

Once this test mode is complete and we have measured ΔA and ϕ_e, we can use this information to properly set α and β to cancel out the errors in the RF section. In general, after correction, Q_{IF} will be

$$Q_{IF} = -\alpha \cdot (1 + \Delta A)\sin(\omega_{IF}t + \varphi(t) + \phi_e) + \beta\cos(\omega_{IF}t + \varphi(t)) \tag{4.19}$$

After some additional manipulation, this expression becomes

$$Q_{IF} = -\alpha\cos(\phi_e) \cdot (1 + \Delta A)\sin(\omega_{IF}t + \varphi(t)) - [\beta + \alpha(1 + \Delta A)\sin(\phi_e)]\cos(\omega_{IF}t + \varphi(t)) \tag{4.20}$$

Ideally we would like

$$\alpha \cos(\phi_e) \cdot (1 + \Delta A) = 1$$

$$\beta + \alpha(1 + \Delta A)\sin(\phi_e) = 0 \qquad (4.21)$$

Solving for α and β gives

$$\alpha = \frac{1}{(1 + \Delta A)\cos(\phi_e)}$$

$$\beta = -\tan(\phi_e) \qquad (4.22)$$

Example 4.1:   Correcting for Amplitude and Phase Imbalance in a Low IF Receiver

A low IF receiver is to receive a 101-MHz RF signal and convert it to a 1-MHz IF using a 100-MHz LO. There are errors in the RF part of the circuit. These need to be corrected within the BBSP part of the radio. Simulate the system and determine the correction coefficients needed.

Solution:
First, a test signal at 101 MHz is fed into the receiver and the IF signals are observed in simulation as shown in Figure 4.9. Clearly these waveforms are not well balanced. The actual amplitude and phase mismatch could be found from this figure, but it can also be found from the baseband signals just as the processor would do. Reading the baseband output data, it is found that: $I_d = -0.445$, $Q_d = 1.488$, $I_{im} = 0.151$, $Q_{im} = -0.0326$. Using (4.17), the amplitude mismatch can be found to be

$$\Delta A = \frac{-2(Q_d Q_{im} + I_d I_{im})}{I_d^2 + Q_d^2} = \frac{-2(1.488 \cdot -0.0326 + -0.445 \cdot 0.151)}{(-0.445)^2 + 1.488^2} = 0.096$$

Using (4.18), the phase error can be found to be

$$\phi_e = 2\tan^{-1}\left[\frac{Q_d I_{im} - I_d Q_{im}}{Q_d^2 + I_d^2}\right] = 2\tan^{-1}\left[\frac{1.488 \cdot 0.151 - 0.445 \cdot 0.0326}{0.445^2 + 1.488^2}\right] = 9.96°$$

Now that the imbalance is known for this radio, compensation can be calculated as

$$\alpha = \frac{1}{(1 + \Delta A)\cos(\phi_e)} = \frac{1}{(1 + 0.096)\cos(9.96°)} = 0.926$$

$$\beta = -\tan(\phi_e) = -\tan(9.96°) = -0.176$$

These numbers are now added to the receiver to correct the signal as shown in Figure 4.8. The simulation is run once more and both I_{im} and Q_{im} are now very close to zero, showing that the correction was successful. This can also be seen by looking at the IF signals again and noting the Q path after correction as seen in Figure 4.10. To test the sideband rejection of this receiver, a signal at 101.1 MHz representing a desired 100-kHz output and a signal at 98.8 MHz representing an undesired 200-kHz output signal was applied to the input. The spectrum of the I_d

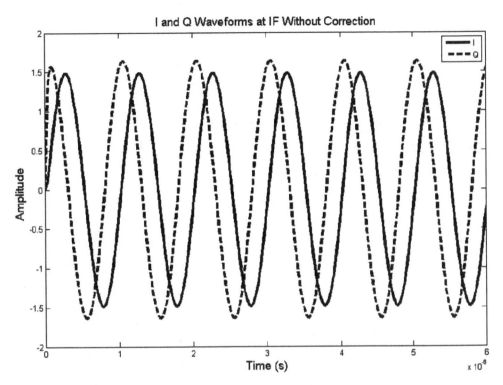

Figure 4.9 I and Q signals of a low IF receiver before correction.

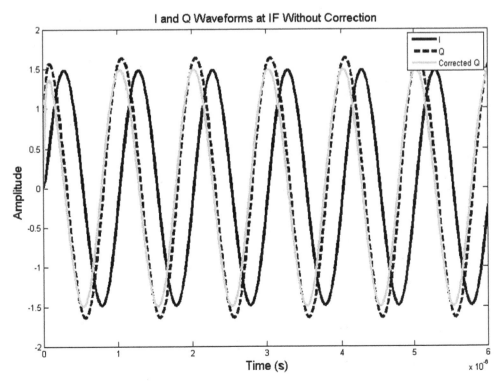

Figure 4.10 IF signals of a low IF receiver including the Q signal after correction is applied.

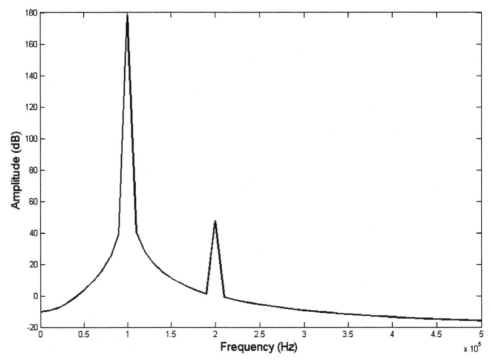

Figure 4.11 Spectrum of the ID signal of a low IF receiver showing rejection of the unwanted sideband.

output is shown in Figure 4.11. This figure shows nearly perfect image rejection. Note that the simulation of this receiver is ideal except for the modeled imperfections explicitly stated in the RF section.

4.6   Sliding IF Transceiver

An architecture that is a compromise between the superheterodyne and the direct-conversion transceiver is called a *sliding IF* architecture shown in Figure 4.12. This

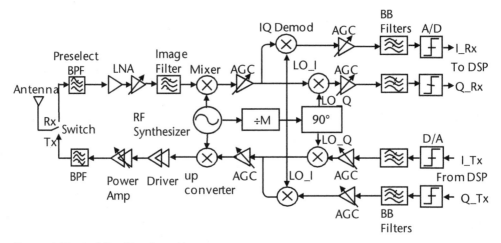

Figure 4.12 A sliding IF radio architecture.

architecture derives the IF LO by dividing the RF LO by some fixed number. As a result, the IF frequency is not fixed, but "walks" in step with a fraction of the frequency of the RF LO. This transceiver with sliding IF still has many of the advantages of the superheterodyne radio (although it is not possible to filter as well at IF), but also removes the need for the extra synthesizer, potentially reducing layout area and power dissipation. Additionally, as there is only one synthesizer, this avoids the possibility of two synthesizer output frequencies mixing together to produce beat frequencies possibly in the baseband.

4.7   An Upconversion-Downconversion Receiver Architecture

A variation on the superheterodyne architecture that can be used when the receiver is expected to handle a wide band of input signals is the upconversion-downconversion architecture shown in Figure 4.13. In this architecture the input signal is first amplified by a broadband LNA and then lowpass filtered to remove noise at the image and then upconverted to a fixed IF frequency above the band of the received signals. It is then filtered to remove the numerous unwanted channels and potentially provide image rejection and then downconverted either to a second lower-frequency IF or directly to baseband. Applications that might make use of this architecture include cable tuners for TV applications or AM radios.

One major advantage of using a dual-conversion architecture is that it eliminates the need for expensive external tracking filters to provide the necessary image rejection at the RF front end. In a lot of cases using a downconverter would mean that for certain channels, the image would be centered on another channel. Thus, placing a filter in the front end that provided image rejection would require that filter to be tunable depending on what channel the receiver was trying to receive. Implementing such filters is difficult using IC technology. Another major problem with downconverting wide bandwidth signals is that harmonics of the synthesizer may fall on other channels, downconverting them on top of the desired channel. Using this architecture moves the images and the VCO harmonics out of the receive band.

Example 4.2:   An Upconversion-Downconversion Receiver for a Cable Tuner Application

In a cable tuner, the RF front end must not only exhibit low noise but it must also have high linearity and be broadband. It must handle up to 135 interfering RF

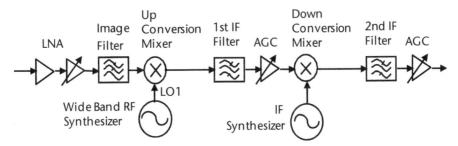

Figure 4.13   System level diagram of an upconversion-downconversion receiver.

channels at +15 dBmV over a frequency range from 47 to 870 MHz. Determine a basic receiver frequency plan to handle this situation.

Solution:
As this is a wideband input, an upconversion-downconversion architecture is chosen. The first IF frequency is chosen to be 1,890 MHz (the European DECT frequency) to eliminate in-band lower-order beat products between the first and second VCO on the chip and to keep the filter cost low due to its high-volume usage. To tune to this IF frequency, a high-side LO is chosen for the wideband RF frequency synthesizer with a range from 1,936 to 2,760 MHz, which places the image frequency range from 3,826 to 4,650 MHz. The front-end filter can easily reject this first image. The main purpose of the RF front end is to convert the incoming band of RF signals to a single IF frequency such that channel selectivity can be achieved by subsequent filtering. After the first IF filter only a few channels will remain and as will be seen in Chapter 5, this will greatly reduce the linearity requirement of the second mixer block. The second LO frequency can be a fixed number. Choosing a low-side injected LO sets the frequency of this synthesizer at 1,840 MHz to provide a second IF frequency of 50 MHz.

4.8   Coherent Versus Noncoherent Receivers

The receivers presented in the last two sections are examples of noncoherent receivers. What this means is that no attempt is made to determine the phase of the transmitted carrier. Because there is no knowledge of the phase, the transmitted carrier will have an arbitrary phase relative to the LO used in the receiver. This creates a problem because if only one mixer were used then the output amplitude would depend on this relative phase difference. If they happened to be in phase, then the output amplitude would be maximized, but if they happened to be phase shifted by 90°, then the output would have a zero amplitude. Thus, two mixers are employed such that the baseband signal can recover both phase and amplitude information as shown in Figure 4.14. By contrast, a coherent radio will recover the phase of the transmitted waveform and remove the need for the second mixer as shown in Figure 4.15. This second type of architecture is not very common as reconstructing the phase of the incoming RF carrier would likely be a very hard circuit to implement.

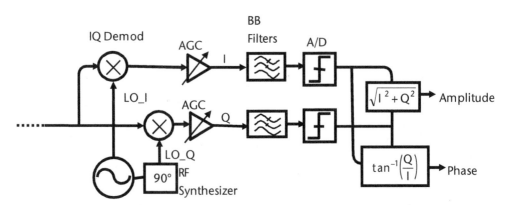

Figure 4.14   A noncoherent receiver producing amplitude and phase information.

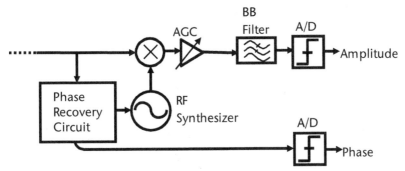

Figure 4.15 A coherent receiver producing amplitude and phase information (not very popular).

In either case listed, there is still the problem of having no reference set to compare against. If the modulation is DPSK, then there is no problem, but the radio in either case, still has the problem of aligning the reference to determine what was sent if a modulation scheme like BPSK or 16QAM is used. So what is typically done in such cases? If an OFDM signal is being used, then likely some carriers will contain reference tones so that both amplitude and phase references can be established. For non-OFDM waveforms, then often at the start of communication, training information will be transmitted to align a local reference to the data. For instance, if the information is encoded as 16QAM, then at the start of transmission there may be a DPSK sequence sent to train the receiver. Once relative phase and amplitude have been established, then the link can gear up into the more complicated data mode.

4.9 Image Rejecting/Sideband Suppression Architectures

Mixing as shown in Figure 4.16(a), always produces two sidebands: one at $\omega_{LO} + \omega_{IF}$ and one at $\omega_{LO} - \omega_{IF}$ caused by multiplying $\cos \omega_{LO}t \times \cos \omega_{IF}t$. It is possible to use a filter after the mixer in the transmitter to get rid of the unwanted sideband for the upconversion case. Similarly, it is possible to use a filter before the mixer in a receiver to eliminate unwanted signals at the image frequency for the downconversion case, as shown in Figure 4.16(b). Alternatively, a single-sideband mixer for the transmit path, or an image-reject mixer for the receive path can be used.

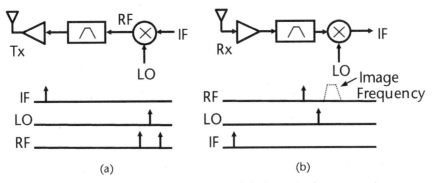

Figure 4.16 (a) Sidebands in upconversion and (b) image in downconversion.

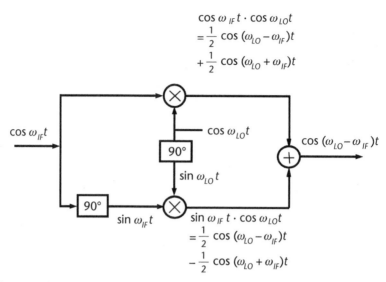

Figure 4.17 A single-sideband mixer. Note that other placements of the phase shifters are possible.

An example of a single-sideband upconversion mixer is shown in Figure 4.17. It consists of two basic mixer circuits, two 90° phase shifters, and a summing stage. As can be shown, the use of the phase shifters and mixers will cause one sideband to add in phase and the other to add in antiphase, leaving only the desired sideband at the output. Which sideband is rejected depends on the placement of the phase shifts or the polarity of the summing block. Other variations of Figure 4.17 are also possible. There are three signals in this circuit: an input, an output, and an LO. A 90° phase shift must be present in two of these signals, but any two of the three signals may be used. Changing the placement of the phase shifters and/or the sign of the phase shift it is easy to select either sideband at the output.

A variant of this circuit, forming an image-reject mixer, used for downconversion is shown in Figure 4.18. In this circuit, at the output, the RF signal adds in phase

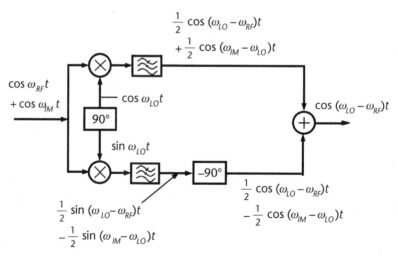

Figure 4.18 A high-side-injected image-reject mixer.

while the image adds in antiphase. Note that two LPFs are added after the mixers to remove the high-frequency summing terms at $\omega_{LO} + \omega_{RF}$ and $\omega_{LO} + \omega_{IM}$.

4.10   An Alternative Single-Sideband Mixer

The image-reject configuration in Figure 4.18 is also known as the *Hartley architecture*. Another possible implementation of an image-reject receiver is known as the *Weaver architecture* shown in Figure 4.19. In this case, the phase shifter after the mixer in Figure 4.18 is replaced by another set of mixers to perform an equivalent operation. The advantage is that all phase shifting takes place only in the LO path and there are no phase shifters in the signal path. As a result, this architecture is less sensitive to amplitude mismatch in the phase-shifting networks and so image rejection is improved. The disadvantage is the additional mixers required, but if the receiver has a two-stage downconversion architecture, then these mixers are already present, so there is no penalty. Note that depending on whether high-side or low-side-injected LOs are chosen, the two phase shifts may have to be changed from 90° to –90° or the summing block might need to be replaced with a subtraction.

4.11   Image Rejection with Amplitude and Phase Mismatch

The ideal requirements are that a phase shift of exactly 90° is generated in the signal path and that the LO has perfect quadrature output signals. In a perfect system, there is also no gain mismatch in the signal paths. In a real circuit implementation, there will be imperfections as shown in Figure 4.20. Therefore, an analysis of how much image rejection can be achieved for a given phase and amplitude mismatch is now performed.

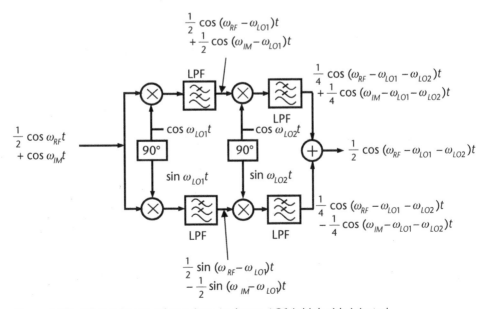

Figure 4.19   Weaver image-reject mixer. As shown, LO1 is high-side injected.

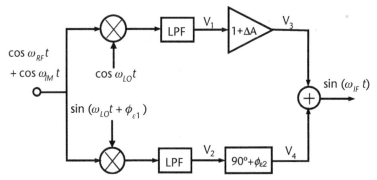

Figure 4.20 Block diagram of an image-reject mixer including phase and gain errors.

The analysis proceeds as follows:

1. The input signal is mixed with the quadrature LO signal through the I and Q mixers to produce signals V_1 and V_2 after filtering. V_1 and V_2 are given by:

$$V_1 = \frac{1}{2}\cos(\omega_{LO} - \omega_{RF})t + \frac{1}{2}\cos(\omega_{IM} - \omega_{LO})t \tag{4.23}$$

$$V_2 = \frac{1}{2}\sin\left[(\omega_{LO} - \omega_{RF})t + \phi_{\varepsilon1}\right] - \frac{1}{2}\sin\left[(\omega_{IM} - \omega_{LO})t - \phi_{\varepsilon1}\right] \tag{4.24}$$

2. Now V_1 experiences an amplitude error relative to V_2 and V_2 experiences a phase shift that is not exactly 90° to give V_3 and V_4, respectively:

$$V_3 = \frac{1}{2}(1+\Delta A)\cos(\omega_{LO} - \omega_{RF})t + \frac{1}{2}(1+\Delta A)\cos(\omega_{IM} - \omega_{LO})t \tag{4.25}$$

$$V_4 = \frac{1}{2}\cos\left[(\omega_{LO} - \omega_{RF})t + \phi_{\varepsilon1} + \phi_{\varepsilon2}\right] - \frac{1}{2}\cos\left[(\omega_{IM} - \omega_{LO})t - \phi_{\varepsilon1} + \phi_{\varepsilon2}\right] \tag{4.26}$$

3. Now V_3 and V_4 are added together. The component of the output due to the RF signal is denoted V_{RF} and is given by

$$\begin{aligned}V_{RF} &= \frac{1}{2}(1+\Delta A)\cos(\omega_{IF}t) + \frac{1}{2}\cos(\omega_{IF}t + \phi_{\varepsilon1} + \phi_{\varepsilon2}) \\ &= \frac{1}{2}(1+\Delta A)\cos(\omega_{IF}t) + \frac{1}{2}\cos(\omega_{IF}t)\cos(\phi_{\varepsilon1} + \phi_{\varepsilon2}) - \frac{1}{2}\sin(\omega_{IF}t)\sin(\phi_{\varepsilon1} + \phi_{\varepsilon2})\end{aligned} \tag{4.27}$$

4. The component due to the image is denoted V_{IM} and is given by

$$V_{IM} = \frac{1}{2}(1+\Delta A)\cos(\omega_{IF}t) - \frac{1}{2}\cos(\omega_{IF}t)\cos(\phi_{\varepsilon2} - \phi_{\varepsilon1}) + \frac{1}{2}\sin(\omega_{IF}t)\sin(\phi_{\varepsilon2} - \phi_{\varepsilon1}) \tag{4.28}$$

5. Only the ratio of the magnitudes is important. The magnitudes are given by

$$|V_{RF}|^2 = \frac{1}{4}\left[\left[\sin\left(\phi_{\varepsilon1} + \phi_{\varepsilon2}\right)\right]^2 + \left[\left(1 + \Delta A\right) + \cos\left(\phi_{\varepsilon1} + \phi_{\varepsilon2}\right)\right]^2\right]$$

$$= \frac{1}{4}\left[1 + \left(1 + \Delta A\right)^2 + 2\left(1 + \Delta A\right)\cos\left(\phi_{\varepsilon1} + \phi_{\varepsilon2}\right)\right] \tag{4.29}$$

$$|V_{IM}|^2 = \frac{1}{4}\left[\left(\sin\left(\phi_{\varepsilon2} - \phi_{\varepsilon1}\right)\right)^2 + \left[\left(1 + \Delta A\right) - \cos\left(\phi_{\varepsilon2} - \phi_{\varepsilon1}\right)\right]^2\right]$$

$$= \frac{1}{4}\left[1 + \left(1 + \Delta A\right)^2 - 2\left(1 + \Delta A\right)\cos\left(\phi_{\varepsilon2} - \phi_{\varepsilon1}\right)\right] \tag{4.30}$$

6. Therefore, the image rejection ratio is given by

$$\text{IRR} = 10\log\frac{|V_{RF}|^2}{|V_{IM}|^2} = 10\log\left\{\frac{1 + (1 + \Delta A)^2 + 2(1 + \Delta A)\cos(\phi_{\varepsilon1} + \phi_{\varepsilon2})}{1 + (1 + \Delta A)^2 - 2(1 + \Delta A)\cos(\phi_{\varepsilon2} - \phi_{\varepsilon1})}\right\} \tag{4.31}$$

If there is no phase imbalance and no amplitude mismatch, then this equation approaches infinity, and so ideally this system will reject the image perfectly. It is only the nonideality of the components that cause finite image rejection. Figure 4.21 shows a contour plot of how much image rejection can be expected for various levels of phase and amplitude mismatch. An amplitude error of about 20% is acceptable for 20 dB of image rejection, but more like 2% is required for 40 dB of image rejection. Likewise, a phase mismatch must be held to less than 1.2° for 40 dB of image rejection, while a phase mismatch of less than 11.4° can be tolerated for 20 dB of image rejection.

4.12   LO Generation

There are many reasons why it may not be desirable to have a synthesizer running at exactly the frequency needed for the mixer. For instance, in full-duplex radios, two synthesizers running at close to the same frequency could pull each other off frequency. As well, direct downconversion radios can have LO feedback reach the antenna and get radiated. If the LO is off frequency, it is necessary to design an LO frequency translation network. One simple way to do this is with integer dividers, which are fairly easy to implement at RF frequencies. Divide-by-two is very popular because a common implementation of this function naturally produces quadrature outputs. Thus, I and Q LO phases are automatically generated. If a more complicated frequency translation is required, then one way to do this is with a network like the one shown in Figure 4.22. This structure uses a single-sideband mixer and a divider. The output frequency of this system will be

$$f_o = f_{syn}\left(\frac{M \pm 1}{M}\right) \tag{4.32}$$

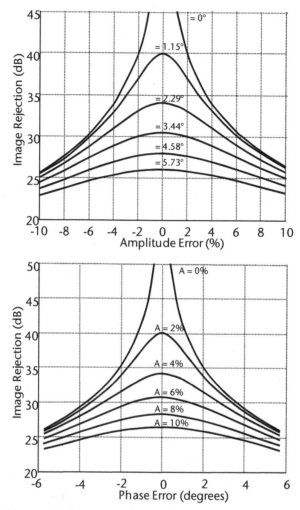

Figure 4.21   Plot of image rejection versus phase and amplitude mismatch.

The choice of the plus or minus in the previous equation depends on which sideband the single-sideband mixer selects.

Example 4.3   Generating a Reference Tone

A GPS receiver that operates at 1.575 GHz is to be integrated with a 2.4-GHz WLAN chip. How can the synthesizer be reused for the GPS receiver?

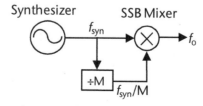

Figure 4.22   Frequency generation network using a single-sideband mixer.

Solution:

If the WLAN synthesizer can be tuned above the band to 2.625 GHz, then using a division ratio of 3/5 will yield the GPS frequency. If we use the divider structure shown in Figure 4.22 with $M = 5$, using the higher-frequency output yields a 6/5 output. Following the output with a divide by two will give an overall output at 3/5 of the input frequency.

Example 4.4:   Generating LOs for an Ultrawideband Radio

An ultrawideband (UWB) radio employs a frequency-hopping spread-spectrum technique. This technique requires the radio to be able to change channels in less than 9 ns. It has been determined that a synthesizer cannot be designed to switch in this length of time. Thus, instead all channels must be derived from a single source tone of 16,896 MHz. Design a LO network that takes the signal at 16,896 MHz and with SSB mixers, MUXs, and dividers generates all these frequencies which correspond to the center of each UWB channel: 3,432 MHz, 3,960 MHz, 4,488 MHz, 5,016 MHz, 5,544 MHz, 6,072 MHz, 6,600 MHz, 7,128 MHz, 7,656 MHz, 8,184 MHz, 8,712 MHz, 9,240 MHz, 9,768 MHz, and 10,296 MHz.

Solution:

Note first that the channels are spaced by 528 MHz. Passing the 16,896-MHz source signal through a series of divide-by-two stages will generate 8,448 MHz, 4,224 MHz, 2,112 MHz, 1,056 MHz, 528 MHz, and 264 MHz. Combinations of these signals can be used to generate all the required channels as shown in Table 4.1, which lists each channel, its associated band group, and the desired LO frequency. Either 8,448 MHz or 4,224 MHz is mixed with either 1,056 MHz or 2,112 MHz in a single-sideband mixer to produce either sum or difference tones. The result of this operation will be 264 MHz from the center of at least one channel. A final single-sideband mixer can combine the result of the first operation with 264 MHz and produce the desired output frequency. An example network to accomplish this is shown in Figure 4.23. Channel switching can now be accomplished in the length of time it takes to change the control on three MUX circuits. This should be a relatively fast operation.

Table 4.1   Summary of a Solution to Generate All the UWB Tones for a Radio

Channel	Band Group	Desired LO Frequency (MHz)	Signals Used to Synthesize
1	1	3,432	4,224–1,056+264
2	1	3,960	4,224–264
3	1	4,488	4,224+264
4	2	5,016	4,224+1,056–264
5	2	5,544	4,224+1,056+264
6	2	6,072	8,448–2,112–264
7	3	6,600	8,448–2,112+264
8	3	7,128	8,448–1,056–264
9	3	7,656	8,448–1,056+264
10	4	8,184	8,448–264
11	4	8,712	8,448+264
12	4	9,240	8,448+1,056–264
13	5	9,768	8,448+1,056+264
14	5	10,296	8,448+2,112–264

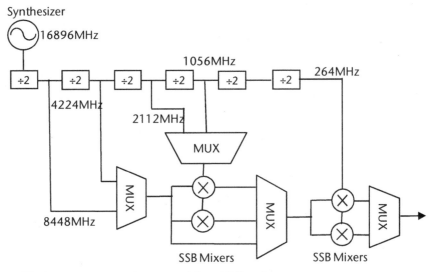

Figure 4.23 Frequency generation network for a UWB radio.

4.13 Channel Selection at RF

All the architectures considered so far in this chapter do not perform any channel selection at RF. In general, at the time of this writing, building tunable bandpass filters at RF frequencies was difficult and rarely employed. However, there are alternative architectures that can be employed in the RF front end to provide channel selection before downconversion [8]. One such architecture is shown in Figure 4.24. Here the RF signal is mixed with an LO centered at the desired channel. The downconverter signal is amplified and applied to a highpass filter. The job of the highpass filter is to remove the desired channel from the overall waveform. Once this is done, the signal is mixed back to the RF frequency and applied to a summing block. Thus, after the summing block, the difference of the incoming signal and the incoming signal's unwanted channels is passed forward. This is a negative feedback loop and therefore through feedback the loop will try to keep the signals at the RF_{out} node as small as possible in any frequency band in which the loop has gain. The corrected RF_{out} signal could then be passed through a more conventional downconversion stage to an IF, or if a direction conversion radio is desired, then the output of the system may be taken at BB_{out} instead.

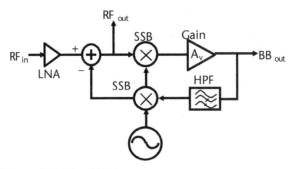

Figure 4.24 An RF channel select architecture.

Assuming that the mixers in Figure 4.24 are ideal, the transfer function of system can be determined. For simplicity, assuming that the gain of the LNA is unity, the transfer function of the system is given by

$$T(s) = \frac{BB_{\text{out}}}{RF_{\text{in}}} = \frac{A_v}{1 + A_v H(s)} \qquad (4.33)$$

where $H(s)$ is the transfer function of the highpass filter. If the highpass filter is assumed to be a simple first-order structure such that

$$H(s) = \frac{s}{s + \omega_o} \qquad (4.34)$$

then (4.33) can be expanded to give

$$T(s) = \frac{\left(\dfrac{A_v}{1 + A_v}\right)(s + \omega_o)}{s + \dfrac{\omega_o}{1 + A_v}} \qquad (4.35)$$

This transfer function contains one zero and one pole. Knowing the bandwidth of the desired signal, ω_o should be chosen to be

$$\omega_o = BW(1 + A_v) \qquad (4.36)$$

Note that at RF the shape of the frequency response would be the same except it would be centered at the LO frequency rather than at DC.

4.14   Transmitter Linearity Techniques

Transmitter linearity and distortion cancellation are an area of great concern in a lot of applications [9, 10]. There are a number of architectural things that can be done to combat these problems. One technique is called feedforward linearization. A typical feedforward architecture is shown in Figure 4.25. Here additional blocks are placed around the RF power amplifier to reduce distortion components. The

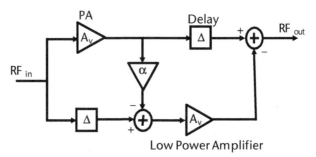

Figure 4.25   A feedforward linearization architecture.

output of the main power amplifier is sensed by an attenuator with a loss of α. Note that α is normally set to be equal to $1/A_v$. The output of this attenuator is subtracted from a delayed version of the input. The delay is included to make sure that the phases are aligned. The idea is that the output of the lower summing block now includes only the distortion components with the main (desired) signal removed. The result is passed through a low-power amplifier and then subtracted from the main output. The gain of the low-power amplifier should match the main power amplifier gain. After the addition, the resulting output RF_{out} should contain the desired signal with no distortion.

While the feedforward technique can work very well, it depends on the gains of the amplifiers to be matched, the low-power amplifier to be linear, and the delays to match the delays through the corresponding amplifiers. If any of these parameters are wrong or incorrectly tuned, the amount of distortion cancellations will decrease proportionally.

Another related technique that does not require so much precision is feedback linearization. However, at RF there is often significant phase shift associated with obtaining the desired gain and this poses stability problems for feedback systems. For this reason, direct RF feedback is not usually employed. Instead, more complicated architectures need to be used such as the one shown in Figure 4.26. Here the output of the PA is sensed and attenuated and then passed through a peak detector. This is compared against the input, which is also passed through a peak detector. The result is amplified and used to control the gain of the transmitter. This technique suffers from the drawback that the relative phases of the input and output are not compared and thus the correction provided by this feedback network is not perfect, but in many cases it is possible to provide a few crucial decibels of needed linearity to a transmitter.

Another feedback loop often used in transmitters is a Cartesian feedback loop. An example of this kind of loop is shown in Figure 4.27. Here the output of the PA is attenuated and passed through a downconverter. Amplifiers then perform the feedback comparison at the IF or baseband rather than at RF. This technique requires more complex circuitry than the previous architecture, but it can provide superior linearization as well.

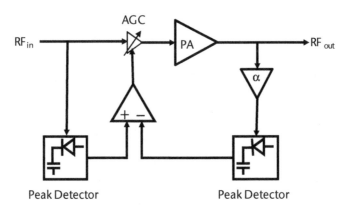

Figure 4.26 A feedback linearization architecture.

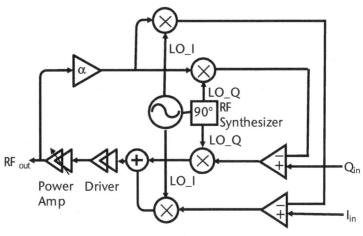

Figure 4.27 A Cartesian feedback linearization architecture.

4.15 Multiple-Input Multiple-Output (MIMO) Radio Architectures

Multiple input multiple output (MIMO) architectures have received increased attention recently [11–13]. The idea of having several radios either at the receiver or at the transmitter or both allows greater transmission range at the same transmitted power level and for the same data rate [14]. Alternatively with MIMO architectures, the system signal-to-noise-ratio (SNR) requirement can be relaxed for a given data rate. Figure 4.28 compares the average SNR per antenna required for a given data rate using different radio configurations for a WLAN link [11]. In Figure 4.28 a conventional single-input-single-output (SISO) 1 × 1 link, namely, one transmitter,

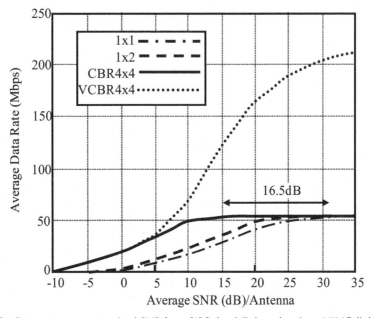

Figure 4.28 Data rate versus required SNR for a SISO 1 × 1 link and various MIMO links including 1 × 2, 4 × 4 CBR, and 4 × 4 VCBR.

one receiver architecture is compared to various MIMO systems such as a 1×2 selection diversity link (spatially separated receiver or transmitter antennas to select the strongest signal), a 4×4 link that uses composite beamforming (CBF) technology and maximal ratio combining, and a vector CBF (VCBF) link. From Figure 4.28 it can be seen that at a data rate of 54 Mbps, the SNR required for a 4×4 link is 16.5 dB lower than that required by a 1×1 link, clearly demonstrating the advantages of MIMO architectures.

When using multiple radios, there is a trade-off between the level of integration and the complexity and yield of the radio ICs. Integrating more radios together leads to more expensive packaging and lower yield, but lower cost and smaller overall designs. However, for example, if a two-radio chip is used and if more than two radios are desired in a particular application, then even two, three, or more dual-radio transceiver chips can be used in a single link, provided that their local oscillators (LO) are all phase synchronized [12]. For instance, two pairs of dual-radio transceiver chips can be used to form a 4×4 MIMO radio system as shown in Figure 4.29. Each transceiver pair consists of a master and a slave dual-radio transceiver chip. The slave chip synchronizes its LO to the master chip by means of the LO porting circuitry, in which the master chip's synthesizer can be used to drive the slave chip's LOs for synchronization.

MIMO transceiver design is a challenge due to the following issues:

1. When multiple radios are integrated in the same die, interference among the transceiver building blocks will be a big concern. In particular, when multi-

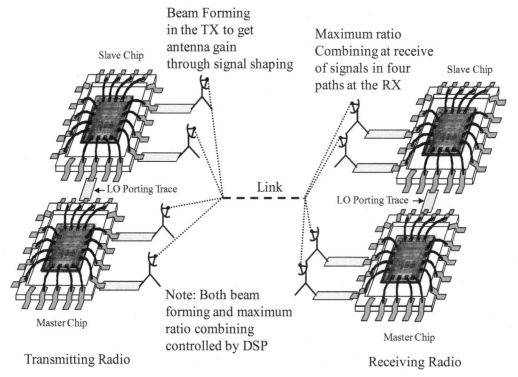

Figure 4.29   A 4×4 MIMO system using two ICs each containing two radios.

band power amplifiers are in operation, their radiation may injection-lock the VCOs and generate cross-talk noise through substrate, package, power supply, or ground. Careful floor planning and proper isolation of the MIMO transceiver layout are critical. The on-chip synthesizer must be carefully designed so that all the VCOs are operating at different frequencies from the PA transmit bands.

2. In a MIMO system, all the LOs in different radio paths need to be synchronized. MIMO calibration requires a loop-back measurement through transmit and receive paths. A loop-back measurement is performed by connecting the transmitter to one antenna and the receiver to the other antenna on the same chip and then matching the gain and phase of both paths. Maximizing the gain from transmitter beamforming and receiver MRC requires matching the gain and phase of each path being combined. If the LO phases of the MIMO radios shift relative to each other, the performance of composite beamforming (CBF) operation will be degraded. However, synchronizing the LOs at high frequencies such as 5.3 GHz for the 802.11a band is not trivial. If more than one PLL is used to generate the LO tones, static phase error and time-variant phase variations among the PLLs will occur. To avoid many of these LO phase drift issues resulting from the use of multiple PLLs, a single frequency synthesizer is typically used to generate all the RF and IF LOs. An optional LO porting block can be added to the architecture, allowing for sharing of a common system RF LO among multiple MIMO chips [12].

3. The isolation between transmitter paths must be as high as possible to maximize the gain from CBF. Finite isolation between transmitters at the PA outputs results in degradation of the maximum possible gain.

4. The isolation between receiver paths must be maximized to maximize the gain from MCR. An isolation of better than 40 dB is often required.

References

[1]   Razavi, B., *RF Microelectronics*, Upper Saddle River, NJ: Prentice Hall, 1998.

[2]   Rogers, J., and C. Plett, *Radio Frequency Integrated Circuit Design*, 2nd ed., Norwood, MA: Artech House, 2010.

[3]   Lee, T. H., *The Design of CMOS Radio Frequency Integrated Circuits*, 2nd ed., Cambridge, U.K.: Cambridge University Press, 2003.

[4]   Crols, J., and M. Steyaert, *CMOS Wireless Transceiver Design*, Dordrecht, the Netherlands: Kluwer Academic Publishers, 1997.

[5]   Gu, Q., *RF System Design of Transceivers for Wireless Communications*, 1st ed., New York: Springer Science and Business Media, 2005.

[6]   Laskar, J., B. Matinpour, and S. Chakradborty, *Modern Receiver Front-Ends*, New York: John Wiley & Sons, 2004.

[7]   Sheng, W., A. Emira, and E. Sanchez-Sinencio, "CMOS RF Receiver System Design: A Systematic Approach," *IEEE Transactions on Circuits and Systems I: Regular Papers*, Vol. 53, No. 5, May 2006, pp. 1023–1034.

[8]   Youssef, S., R. van der Zee, and B. Nauta, "Active Feedback Receiver with Integrated Tunable RF Channel Selectivity, Distortion Cancelling, 48dB Stopband Rejection and >+12dBm Wideband IIP3, Occupying <0.06mm^2 in 65nm CMOS," *Proc. International Solid-State Circuits Conference*, 2012, pp. 166–167.

[9]   Cripps, S. C., *RF Power Amplifiers for Wireless Communications*, 2nd ed., Norwood, MA: Artech House, 2006.

[10] Pothecary, N., *Feedforward Linear Power Amplifiers*, Norwood, MA: Artech House, 1999.

[11] Rahn, D. G., et al., "A Fully Integrated Multi-Band MIMO WLAN Transceiver RFIC," *IEEE Journal of Solid State Circuits*, Vol. 40, No. 8, August 2005, pp. 1629–1641.

[12] Rogers, J. W. M., et al., "A Multi-Band $\Delta\Sigma$ Fractional-N Frequency Synthesizer for a MIMO WLAN Transceiver RFIC," *Journal of Solid State Circuits*, Vol. 40, No. 3, March 2005, pp. 678–689.

[13] Sugar, A. G., R. M. Masucci, and D. Rahn, Cognio, Inc., "Multiple-Input Multiple-Output Radio Transceiver," United States Patent No. 6,728,517, April 27, 2004.

[14] Murch, R. D., and K. B. Letaief, "Antenna Systems for Broadband Wireless Access," *IEEE Communications Magazine*, April 2002, pp. 76–83.

CHAPTER 5

System-Level Design Considerations

5.1 Introduction

This chapter will discuss some of the most important design considerations when specifying the requirements of all the components in the system. The idea is to provide detailed equations and calculations that allow the system's designer to understand the impact of block level performance specifications on the overall system level performance. For additional information the reader should consider [1–10].

5.2 The Noise Figure of Components in Series

Once the noise figures of all the components in a receiver chain are known, it is necessary to compute the total overall noise figure of the chain. For components in series, as shown in Figure 5.1, one can calculate the total output noise $N_{o(total)}$ and output noise due to the source $N_{o(source)}$ to determine the noise figure assuming all components are matched to a defined impedance.

The output signal S_o is given by

$$S_o = S_i \cdot G_1 \cdot G_2 \cdot G_3 \tag{5.1}$$

The input noise is

$$N_{i(source)} = kT \tag{5.2}$$

The total output noise is

$$N_{o(total)} = N_{i(source)}G_1G_2G_3 + N_{o1(added)}G_2G_3 + N_{o2(added)}G_3 + N_{o3(added)} \tag{5.3}$$

The output noise due to the source is

$$N_{o(source)} = N_{i(source)}G_1G_2G_3 \tag{5.4}$$

Finally, the noise factor can be determined as

$$F = \frac{N_{o(total)}}{N_{o(source)}} = 1 + \frac{N_{o1(added)}}{N_{i(source)}G_1} + \frac{N_{o2(added)}}{N_{i(source)}G_1G_2} + \frac{N_{o3(added)}}{N_{i(source)}G_1G_2G_3}$$

$$= F_1 + \frac{F_2 - 1}{G_1} + \frac{F_3 - 1}{G_1G_2} \tag{5.5}$$

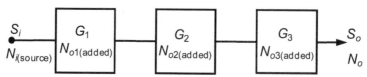

Figure 5.1   Noise figure of cascaded circuits with gain and noise added shown in each.

The above formula shows how the presence of gain preceding a stage causes the effective noise figure to be reduced compared to the measured noise figure of a stage by itself. For this reason, we typically design systems with a low noise amplifier at the front of the system. We note that the noise figure of each block is typically determined for the case in which a standard input source (e.g., 50 Ω) is connected. The formula in (5.5) can also be used to derive an equivalent model of each block as shown in Figure 5.2. If the input noise power when measuring noise figure is

$$N_{i(source)} = kT \tag{5.6}$$

and noting from the manipulation of (3.14) that

$$N_{o(added)} = (F-1)N_{o(source)} \tag{5.7}$$

Now dividing both sides of (5.7) by G

$$N_{i(added)} = (F-1)\frac{N_{o(source)}}{G} = (F-1)N_{i(source)} = (F-1)kT \tag{5.8}$$

Thus, the input referred noise model for cascaded stages as shown in Figure 5.2 can be derived.

Example 5.1:   Cascaded Noise Figure and Sensitivity Calculation

Find the effective noise figure and the noise floor of the system shown in Figure 5.3. The system consists of a filter with 3-dB loss, followed by a switch with 1-dB loss, an LNA, and a mixer. Assume the system needs an SNR of 7 dB for a bit error rate of 10^{-3}. Also, assume that the system bandwidth is 200 kHz.

Solution:
Since the bandwidth of the system has been given as 200 kHz, the noise floor of the system can be determined:

$$\text{Noise Floor} = -174 \text{ dBm} + 10 \log_{10}(200{,}000) = -121 \text{ dBm}$$

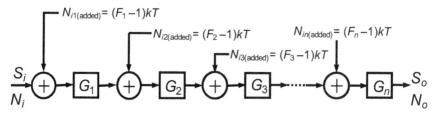

Figure 5.2   Equivalent noise model of a circuit.

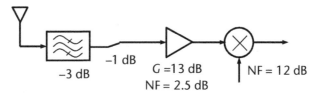

Figure 5.3   System for performance calculation.

We make use of the cascaded noise figure equation and determine that the overall system noise figure is given by

$$NF_{TOTAL} = 3\text{ dB} + 1\text{ dB} + 10\log_{10}\left[10^{2.5/10} + \frac{10^{12/10} - 1}{10^{13/10}}\right]$$

$$= 3\text{ dB} + 1\text{ dB} + 10\log_{10}\left[1.78 + \frac{15.84 - 1}{20}\right] \approx 8\text{ dB}$$

Note that if the mixer also has gain, then possibly the noise due to the IF stage may be ignored. In a real system this would have to be checked, but here we will ignore noise in the IF stage.

As it was stated that the system requires an SNR of 7 dB, the sensitivity (the minimum power level that can be successfully detected) of the system can now be determined:

$$\text{Sensitivity} = -121\text{ dBm} + 7\text{ dB} + 8\text{ dB} = -106\text{ dBm}$$

Thus, the smallest allowable input signal is –106 dBm. If this is not adequate for a given application, then a number of things can be done to improve this:

1. A smaller bandwidth could be used. However, the signal bandwidth is usually standardized and so cannot be changed.
2. The loss in the preselect filter or switch could be reduced. For example, the LNA could be placed in front of one or both of these components.
3. The noise figure of the LNA could be improved.
4. The LNA gain could be increased reducing the effect of the mixer on the system NF.
5. A lower NF in the mixer would also improve the system NF.
6. If a lower SNR for the required BER could be tolerated, then this would also help.

In a modern completely integrated radio, often only the interface with off chip circuitry has a standard impedance and the interfaces between blocks on chip tend to be undefined. In this case defining a noise figure for each sub-block is less convenient. On chip most blocks may be better characterized by a voltage gain G_{vi} (defined as the output voltage divided by the input voltage with the intended load impedance attached to the circuit) and an added noise voltage. In this case the overall input referred noise for the system can be given as

$$v_{ni(added)}^2 = v_{n1}^2 + \frac{v_{n2}^2}{G_{v1}^2} + \frac{v_{n3}^2}{G_{v1}^2 G_{v2}^2} + \dots \tag{5.9}$$

where v_{n1}, v_{n2}, and v_{n3} are the input referred noise voltages of the three stages. If we assume that the input of the first block is matched to the source resistance, then the noise figure is given by [8]

$$F = \frac{N_{i(total)}}{N_{i(source)}} = \frac{\dfrac{v_{ni(added)}^2 + v_{ni(source)}^2}{R_s}}{\dfrac{v_{ni(source)}^2}{R_s}} = 1 + \frac{v_{n1}^2 + \dfrac{v_{n2}^2}{G_{v1}^2} + \dfrac{v_{n3}^2}{G_{v1}^2 G_{v2}^2} + \dots}{v_{ni(source)}^2}$$

$$\tag{5.10}$$

$$= 1 + \frac{v_{n1}^2 + \dfrac{v_{n2}^2}{G_{v1}^2} + \dfrac{v_{n3}^2}{G_{v1}^2 G_{v2}^2} + \dots}{kTR_s} = 1 + \frac{v_{n1}^2}{kTR_s} + \frac{v_{n2}^2}{kTR_s G_{v1}^2} + \frac{v_{n3}^2}{kTR_s G_{v1}^2 G_{v2}^2} + \dots$$

This formula would provide the system designer with the opportunity to specify voltage gain and input referred noise voltage rather than noise figure which makes more sense.

Example 5.2:   On-Chip NF Confusion

A system-level designer unfamiliar with IC design has set the system level requirements for a radio. The system designer has made the mistake of assuming that all blocks are matched to 50 Ω. The RF front end has a gain of 10 dB and an NF of 4 dB. The first block in the IF stage is a variable gain amplifier (VGA). The system-level designer specifies this block to have an NF of 10 dB so that the overall NF of the LNA, mixer and VGA will be

$$F = F_{LNA+Mix} + \frac{F_{VGA} - 1}{G_{LNA+Mix}} = 10^{0.4} + \frac{10 - 1}{10} = 3.41 = 5.33 \text{ dB}$$

This is below the target for this design of 6.5 dB and makes sure that the VGA is not a dominant concern. Once design is underway, the mixer designer adds a buffer with an output impedance of 5 Ω to drive the VGA which has an input impedance of 5 kΩ. The VGA designer simulates his block using a 50 Ω source and finds he has an NF of 15 dB. Assuming the VGA input can be modeled with the 5 kΩ input resistance parallel to input noise current, does he meet the intended noise requirement for the block?

Solution:
Appling the same formula as the system level designer initially used, the noise figure of the system with a VGA with a 15-dB noise figure would be

$$F = F_{LNA+Mix} + \frac{F_{VGA} - 1}{G_{LNA+Mix}} = 2.51 + \frac{31.6 - 1}{10} = 5.57 = 7.46 \text{ dB}$$

This is about 1 dB over the specification, so clearly a concern for the system designer. The noise voltage at the VGA input with a 50 Ω input is

$$N_{VGA} = \frac{v_{nVGA}^2}{R_s} = (F-1)kT \Rightarrow v_{nVGA} = \sqrt{(F-1)kTR_S} = \sqrt{\frac{30.6}{4} \times 4kTR_S}$$

$$= 2.76 \times 0.9\, nV = 2.49\, nV/\sqrt{Hz}$$

If the same noise current is now fed into $5\,\Omega\|5\,k\Omega$ instead of $50\,\Omega\|5\,k\Omega$, the noise voltage will be approximately 1/10 because in either case the 5 kΩ resistor can be ignored. So using $V_{nVGA} = 0.249$ nV results in:

$$F = F_{RF} + \frac{v_{nVGA}^2}{kTR_sG_{vRF}^2} = 2.51 + \frac{4 \times (0.249nV)^2}{(0.9nV)^2 \times 10} = 2.51 + 0.03 = 2.54 = 4.05\, dB$$

So the VGA has nearly no effect on noise, but a real VGA would have both current and voltage noise, leading to somewhat higher noise. Also, noise due to the extra buffer is low due to the G_{RF} preceding it. Note that G_{RF} is power gain and G_{vRF} is voltage gain. So in this case we see that even though the block looks like it fails the specification, it actually only adds a small fraction of the noise allowed by the original specification. However, because 15 dB is more than the 10 dB specified, this now results in pressure on the VGA designer by the system designer, possibly resulting in wrongful termination and a series of lawsuits. The systems designer should have used (5.10). It should be noted that the model of the noise as coming from a 5-kΩ resistor at the input of the VGA is crucial for this example. More realistically, there would also be a series component in which case a lower driving impedance would not have been as beneficial. In fact, if the noise model consisted only of a series resistor, then a higher driving impedance would have helped to reduce the noise. However, both cases show that if the impedance is not matched to 50 Ω, simply applying the standard equations may not result in the correct noise figure.

5.3   The Linearity of Components in Series

Linearity can also be computed for components in series. Starting with two amplifiers in series that have unique nonlinear transfer functions of

$$v_{o1} = k_{a1}v_i + k_{a2}v_i^2 + k_{a3}v_i^3$$

$$v_{o2} = k_{b1}v_{o1} + k_{b2}v_{o1}^2 + k_{b3}v_{o1}^3 \tag{5.11}$$

each block will have an IIP3 of

$$v_{IIP3\_1} = 2\sqrt{\frac{k_{a1}}{3|k_{a3}|}}$$

$$v_{IIP3\_2} = 2\sqrt{\frac{k_{b1}}{3|k_{b3}|}} \tag{5.12}$$

From (5.11) v_{o2} can be expanded to give an overall transfer function of

$$v_{o2} = k_{a1}k_{b1}v_i + \left(k_{a2}k_{b1} + k_{a1}^2k_{b2}\right)v_i^2 + \left(k_{b1}k_{a3} + 2k_{b2}k_{a1}k_{a2} + k_{b3}k_{a1}^3\right)v_i^3 + \ldots \tag{5.13}$$

Note that this equation is truncated after the third-order terms and it is assumed that higher-order terms are less important.

Now applying the definition of IIP3 to the overall transfer function yields

$$v_{\text{IIP3}} = 2\sqrt{\frac{k_{a1}k_{b1}}{3\left|k_{b1}k_{a3} + 2k_{b2}k_{a1}k_{a2} + k_{b3}k_{a1}^3\right|}} \tag{5.14}$$

Now if both sides are squared and inverted, this becomes

$$\frac{1}{v_{\text{IIP3}}^2} = \frac{3}{4} \cdot \frac{\left|k_{b1}k_{a3}\right| + \left|2k_{b2}k_{a1}k_{a2}\right| + \left|k_{b3}k_{a1}^3\right|}{k_{a1}k_{b1}}$$

$$= \frac{3}{4} \cdot \left(\frac{\left|k_{b1}k_{a3}\right|}{k_{a1}k_{b1}} + \frac{\left|2k_{b2}k_{a1}k_{a2}\right|}{k_{a1}k_{b1}} + \frac{\left|k_{b3}k_{a1}^3\right|}{k_{a1}k_{b1}}\right) \tag{5.15}$$

If it is assumed that the second-order terms k_{a2} and k_{b2} are small compared to the first-order terms, this expression can be simplified:

$$\frac{1}{v_{\text{IIP3}}^2} = \frac{1}{v_{\text{IIP3\_1}}^2} + \frac{3k_{b2}k_{a2}}{2k_{b1}} + \frac{k_{a1}^2}{v_{\text{IIP3\_2}}^2} \approx \frac{1}{v_{\text{IIP3\_1}}^2} + \frac{k_{a1}^2}{v_{\text{IIP3\_2}}^2} \tag{5.16}$$

In general, the IIP3 of cascaded stages with voltage gain A_v is given by

$$\frac{1}{v_{\text{IIP3}}^2} = \frac{1}{v_{\text{IIP3\_1}}^2} + \frac{A_{v1}^2}{v_{\text{IIP3\_2}}^2} + \frac{A_{v1}^2 A_{v2}^2}{v_{\text{IIP3\_3}}^2} + \dots \tag{5.17}$$

Because (5.17) uses the square of the voltage, assuming all blocks are matched to 50 Ω, it can be rewritten in terms of power:

$$\frac{1}{IIP3} = \frac{1}{IIP3\_1} + \frac{G_1}{IIP3\_2} + \frac{G_1 G_2}{IIP3\_3} + \dots \tag{5.18}$$

where G_1 and G_2 are the power gains of the respective blocks.

Similarly, the 1dB compression point of a cascaded system would be

$$\frac{1}{v_{\text{1dB}}^2} = \frac{1}{v_{\text{1dB\_1}}^2} + \frac{A_{v1}^2}{v_{\text{1dB\_2}}^2} + \frac{A_{v1}^2 A_{v2}^2}{v_{\text{1dB\_3}}^2} + \dots \tag{5.19}$$

IIP2 for series blocks has a slightly different form. Using the two-element system discussed above gives

$$v_{\text{IIP2}} = \frac{k_{a1}k_{b1}}{k_{a2}k_{b1} + k_{a1}^2 k_{b2}} \tag{5.20}$$

Inverting this equation gives

$$\frac{1}{v_{\text{IIP2}}} = \frac{k_{a2}k_{b1} + k_{a1}^2 k_{b2}}{k_{a1}k_{b1}} = \frac{k_{a2}}{k_{a1}} + \frac{k_{a1}k_{b2}}{k_{b1}} = \frac{1}{v_{\text{IP2\_1}}} + \frac{k_{a1}}{v_{\text{IP2\_2}}} \tag{5.21}$$

In general, the IIP2 of cascaded stages will be given by

$$\frac{1}{v_{\text{IIP2}}} = \frac{1}{v_{\text{IIP2\_1}}} + \frac{A_{v1}}{v_{\text{IIP2\_2}}} + \frac{A_{v1} A_{v2}}{v_{\text{IIP2\_3}}} + \dots \tag{5.22}$$

Note that, unlike noise figure, nonlinearity is usually dominated by the later stages of the system. One should also take care when applying the above formulas and results to actual cascaded networks. These results were derived assuming no filtering and therefore the gain of each frequency component experiences the same gain through the system. If filters are present, they need to be carefully considered and results may vary greatly from what is predicted by blindly applying these formulas.

5.4   Dynamic Range

So far, we have discussed noise and linearity in circuits. Noise determines how small a signal a receiver can handle, while linearity determines how large a signal a receiver can handle. If operation up to the 1dB compression point is allowed (for about 10% distortion, or IM3 is about –20 dB with respect to the desired output), then the dynamic range is from the minimum detectable signal to this point. This is illustrated in Figure 5.4. In this figure, intermodulation components are above the minimum detectable signal for $P_{in} > -32$ dBm, for which $P_{out} = -23$ dBm. Thus, for any P_{out} between the minimum detectable signal of –105 dBm and –23 dBm, no intermodulation components can be seen and the spurious free dynamic range is 82 dB.

Example 5.3:   Determining Dynamic Range

In Example 3.1 we determined the sensitivity of a receiver system. Figure 5.5 shows this receiver again with the linearity of the mixer and LNA specified. Determine the dynamic range of this receiver.

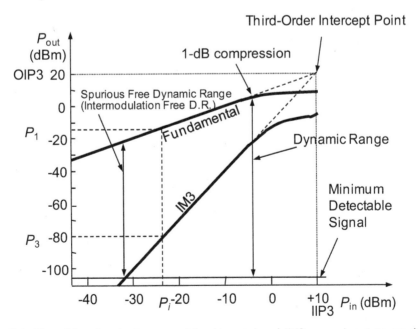

Figure 5.4 Plot of input output power of fundamental and IM3 versus input power showing dynamic range.

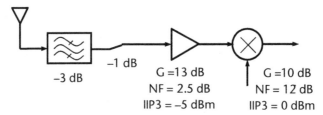

Figure 5.5   System example circuit.

Solution:
The overall receiver has a gain of 19 dB. The minimum detectable signal from Example 3.1 is –106 dBm or –87 dBm at the output. The IIP3 of the LNA and mixer combination from (5.18) is

$$\frac{1}{\text{IIP3}} = \frac{1}{316.2\,\mu\text{W}} + \frac{20}{1m\text{W}} = 2.316 \cdot 10^4$$

$$\text{IIP3} = 43.2\,\mu\text{W} \Rightarrow -13.6 \text{ dBm}$$

This referred to the input is –13.6 + 4 = –9.6 dBm. The IIP3 of the mixer by itself (with a perfect LNA) referred to the input would be 0 –13 + 4 = –9 dBm. This is close to the IIP3 with the real LNA, therefore, the mixer dominates the IIP3 for the receiver as expected. The 1 dB compression point will be 9.6 dB lower than the –9.6 dBm calculated with the real LNA, or –19.2 dBm. Thus, the dynamic range of the system will be –19.2 + 106 = 86.8 dB.

Example 5.4:   Effect of Bandwidth on Dynamic Range

The data transfer rate of the previous receiver can be greatly improved if we use a bandwidth of 80 MHz rather than 200 kHz. What does this do to the dynamic range of the receiver?

Solution:
This system is the same as the last one except that now the bandwidth is 80 MHz. Thus, the noise floor is now

$$\text{Noise Floor} = -174 \text{ dBm} + 10 \log_{10}(80 \cdot 10^6) = -95 \text{ dBm}$$

Assuming that the same SNR is required:

$$\text{Sensitivity} = -95 \text{ dBm} + 7 \text{ dB} + 8 \text{ dB} = -80 \text{ dBm}$$

Thus, the dynamic range is now –19.2 + 80 = 60.8 dB. To get this back to the value in the previous system, we would need to increase the linearity of the receiver by 26 dB. This would be no easy task.

5.5   Image Signals and Image Reject Filtering

At the RF frequency there are filters to remove out-of-band signals that may be picked up by the antenna. Any filter in the RF section of the radio must be wide enough to pass the entire receive band, and therefore can do little about any in-band interferers. In a superheterodyne receiver, the filters in the RF section also have the added task of removing the image frequency and are thus sometimes called image reject filters. A superheterodyne receiver takes the desired RF input signal and mixes it with some reference signal to extract the difference frequency. The LO reference is mixed with the input to produce a signal at the difference frequency of the LO and RF as shown in Figure 5.6. As mentioned earlier, a signal on the other side of the LO at the same distance from the LO will also mix down "on top" of the desired frequency. Thus, before mixing can take place, this unwanted image frequency must be removed.

Thus, another important specification in a receiver is how much image rejection it has. Image rejection is defined as the ratio of the gain G_{sig} of the desired signal to the gain of the image signal G_{im}:

$$IR = 10\log\left(\frac{G_{sig}}{G_{im}}\right) \tag{5.23}$$

In general, a receiver must have an image rejection large enough so that in the case of the largest possible image signal and the weakest receive channel power, the ratio of the channel power to the image power, once downconverted, is still larger than the minimum required SNR for the radio. This is illustrated in Figure 5.6.

The amount of filtering can be calculated by knowing the undesired frequency with respect to the filter center frequency, the filter bandwidth, and the filter order. The following equation can be used for this calculation

$$A_{dB} = \frac{n}{2} \times 20\log\left(\frac{f_{ud} - f_c}{f_{be} - f_c}\right) = \frac{n}{2} \times 20\log\left(2\frac{\Delta f}{f_{BW}}\right) \tag{5.24}$$

where A_{dB} is the attenuation in decibels, n is the filter order, and thus $n/2$ is the effective order on each edge, f_{ud} is the frequency of the undesired signal, f_c is the filter center frequency, and f_{be} is the filter band edge.

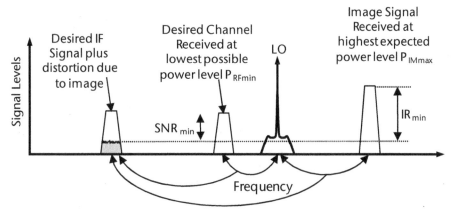

Figure 5.6   Illustration of how the required image rejection for a receiver is determined.

Example 5.5:   Image Reject Filtering

A system has an RF band from 902–928 MHz and a 200-kHz channel bandwidth and channel spacing. The first IF is at 70 MHz and the RF section contains a filter with a 26-MHz bandwidth. Determine the order of filter required to get a worst-case image rejection of better than 50 dB. If the received image signal power is –40 dBm, the minimum input signal power is –75 dBm, and the required SNR for this modulation is 9.5 dB, will this be enough image rejection?

Solution:

The frequency spectrum is shown in Figure 5.7. At RF, the local oscillator frequency f_{LO} is tuned to be 70 MHz above the desired RF signal so that the desired signal will be mixed down to IF at 70 MHz. Thus, f_{LO} is adjustable between 972 MHz and 998 MHz to allow signals between 902 MHz and 928 MHz to be received. The image occurs 70 MHz above f_{LO}. The worst case will be when the image frequency is closest to the filter frequency. This occurs when the input is at 902 MHz, the LO is at 972 MHz and the image is 1,042 MHz. The required filter order n can be calculated by solving (5.24) using f_{BW} = 26 MHz and Δf = 70 + 44 + 13 = 127 MHz as follows:

$$n = \frac{2 \times A_{dB}}{20 \times \log(2\Delta f / f_{BW})} = 5.05$$

As the order should be an even number, a sixth-order filter is used and total attenuation is calculated to be 59.4 dB.

Now with this filter specification, the –40-dBm image signal will be attenuated by 50 dB so after the filter, this signal power will be –90 dBm. This will be mixed on top of the desired signal which has a minimum power level of –75 dBm, giving a signal to distortion ratio of –75 dBm – (–90 dBm) = 15 dB. This is higher than the required SNR of 9.5 dB so this should be enough image rejection.

Example 5.6:   Determining the Required Image Rejection in a QPSK Receiver

Suppose a QPSK receiver needs to have a BER of 1% or better. The RF signal is 1,000 Hz and the IF is chosen to be 100 Hz. The data rate is 1 symbol/s. The LO is low-side injected at 900 Hz. The receiver is modeled in Simulink and a simplified schematic is shown in Figure 5.8. Here no nonlinearity in the signal path is modeled, and an IF filter is implemented as a sixth-order Butterworth filter with a bandwidth of 90–110 Hz. The mixers are ideal multipliers. At baseband the I and

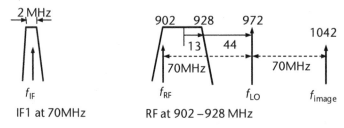

Figure 5.7   Signal spectrum for filter example.

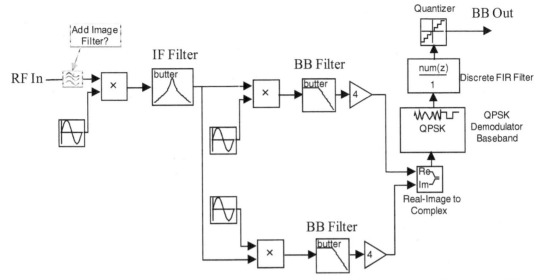

Figure 5.8 A Simulink model of a superheterodyne receiver. Should an image filer be added? If so what qualities should it have?

Q paths are filtered using a sixth-order Butterworth filter with a bandwidth of 1 Hz and the signals are then recombined using a real/imaginary to complex number block and passed into an QPSK demodulator. The output of the demodulator is sampled 10 times over half a symbol period using the FIR filter block and then the average (which may not be an integer) is passed through an ideal quantizer to make sure the output takes on the value of the closest symbol. The phase of the received signal is aligned with the transmitted data manually by adjusting the phase of the IF LO signals. The output sampling time is also adjusted manually.

To verify the functionality of this receiver, a signal was applied to the input with varying SNR levels and the BER rate was computed. The BER can be determined easily by simply comparing the bits produced by the receiver with what was sent by the transmitter. Table 5.1 compares the simulated BER to one predicted from theory, which states that the BER of a QPSK receiver is:

$$P_B\left(\frac{E_s}{N_0}\right) \approx \frac{2}{\log_2 4} \cdot Q\left(\sqrt{2\frac{E_s}{N_0}} \cdot \sin\left(\frac{\pi}{4}\right)\right)$$

where E_s/N_0 is the energy per symbol to the noise density ratio here assumed to be equal to the SNR for the RF section of the radio and the function $Q(x)$ is the is the area underneath the tail of a Gaussian probability density function. Note that

Table 5.1   Results of BER Simulations

SNR	Theoretical BER	Simulated BER
5 dB	3.8%	3.65%
6 dB	2.33%	2.33%
7 dB	1.28%	1.25%
8 dB	0.61%	0.48%
9 dB	0.25%	0.18%
10 dB	0.08%	0.07%

the results compare very favorably with theory. Small discrepancies may be due to the lack of a match filter and the fact that the bandwidth was approximated as the channel width of 2 Hz in this simple example.

The received RF spectrum and I and Q baseband signals are plotted in Figure 5.9 for a SNR of 8 dB. The BER rate at this level was 0.48%, which will be close to the minimum SNR that can be tolerated by this radio and still achieve the desired 1% BER. Now if another signal is received at the image frequency of 800 Hz, which is 50 dB higher than the desired signal, then the image filter shown in Figure 5.8 will be needed. How much image rejection is required to maintain a BER of 1% or better?

Solution:
With the image tone present, the RF spectrum can be expected to look like Figure 5.10. Using the receiver as it was previously designed yields a BER of 62.4%,

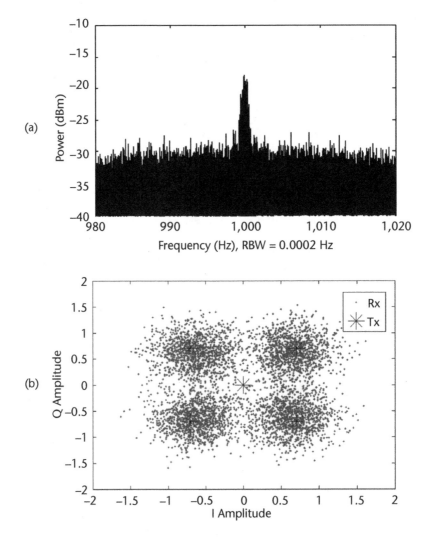

Figure 5.9 (a) Received RF spectrum plot of a RF signal with an SNR of 8 dB and (b) recovered baseband constellation of 5,000 symbols with an SNR of 8 dB.

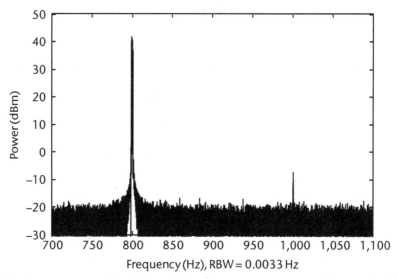

Figure 5.10   Frequency plot of the RF spectrum with a strong image tone about 50 dB higher than the desired signal.

which means that the receiver is not functioning. In order for the receiver to function as before, we need to attenuate the image power until it is below the noise floor, which is 10 dB below the signal power; thus we require 60 dB of image rejection (the signal is presently 50 dB higher than the RF signal and it needs to become at least 10 dB smaller than the RF signal) to provide an equivalent SNR to what we previously had. Let us assume for this example that there are 25 channels in this radio between 975 Hz and 1,025 Hz for a total RF bandwidth of 50 Hz. Then the order of the filter can be computed as follows:

$$n = \frac{2A_{dB}}{20\log\left(\dfrac{f_{ud} - f_c}{f_{be} - f_c}\right)} = \frac{2(60)}{20\log\left(\dfrac{800 - 1000}{975 - 1000}\right)} = 6.6$$

Therefore, from the previous calculation, we will choose eighth order (an even number) for our filter. We place this filter into the signal path and repeat the simulation with an SNR of 8 dB. After this was done, the BER became 0.53% for 2,000 symbols transmitted, which is very close to the 0.48% obtained without the image signal. Results are shown in Figure 5.11. For comparison a sixth-order filter was also simulated. In this case the BER increased to 4.75% which is above the 1% desired. This shows that this filter would not be sufficient as the simple calculation predicted. Results for this simulation are shown in Figure 5.12.

5.6   Blockers and Blocker Filtering

The receiver must be able to maintain operation and to detect the desired signal in the presence of other signals, often referred to as *blockers*. These other signals could be of large amplitude and could be close by in frequency for example they could be signals in other channels of the same radio band. This is an example of the near-far

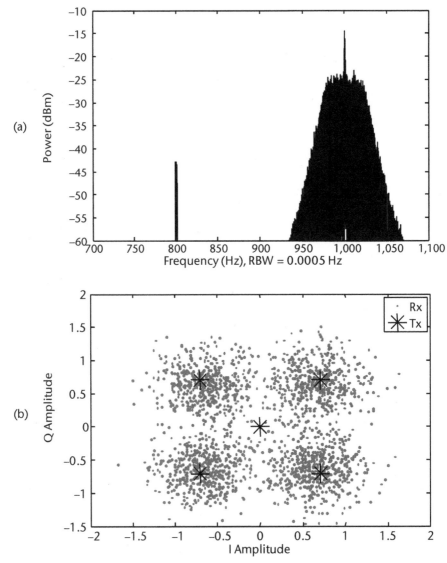

Figure 5.11 (a) RF spectrum after filtering with an eighth-order bandpass image reject filter, and (b) IQ constellation of received symbols with an eighth-order image filter.

problem that occurs when the desired signal is far away and one or more interfering signals are close by and hence, much larger than the wanted signal. A large blocker must not saturate the radio and therefore the 1-dB compression point of the radio must be higher than the blocker power level to avoid saturating the radio.

The intermodulation products of blockers can also be a very big problem. Consider the case where a desired channel is detected at its minimum power level. Two close by channels are also received at their maximum receive power. If these signals are at frequencies such that their IM3 products fall on top of the desired signal they will act to reduce the SNR of the desired channel causing an increase in bit error rate. Therefore, the circuits in the radio must have a sufficiently high linearity so that this does not happen. Once the received band is downconverted to an IF or

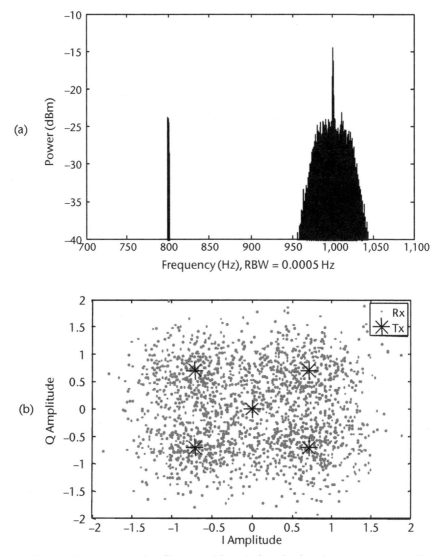

Figure 5.12 (a) RF spectrum after filtering with a sixth-order bandpass image reject filter, and (b) IQ constellation of received symbols with a sixth-order image filter.

baseband frequency, filters may be added with a passband narrower than the whole radio band. As a result, strong adjacent channel signals are filtered and this will reduce the linearity requirements of blocks after the filter.

Example 5.7: How Blockers Are Used to Determine Linearity

Consider a typical blocker specification for a receiver shown in Figure 5.13(a). The input signal is at –102 dBm and the required SNR, with some safety margin, is 11 dB. Calculate the required input linearity of the receiver. If the receiver front end has 20 dB of gain and is followed by an IF filter with 30 dB of attenuation at offsets between 500 kHz and 2.5 MHz and an attenuation of 50 dB at offsets greater than 2.5 MHz as shown in Figure 5.13(b), what is the required linearity of the first circuit in the IF stage?

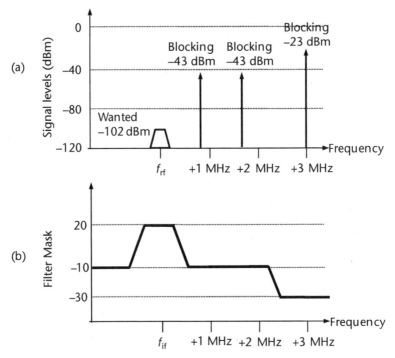

Figure 5.13 (a) Minimum detectable signal and blocker levels at the RF input of the radio, and (b) IF filter mask to help reduce the effect of the blockers.

Solution:

With nonlinearity, third-order intermodulation between the pair of blockers will cause interference directly on top of the signal. The level of this disturbance must be low enough so that the signal can still be detected. The other potential problem is that the large blocker at –23 dBm can cause the amplifier to saturate, rendering the amplifier helpless to respond to the desired signal, which is much smaller. In other words, the receiver has been blocked.

The blocker inputs at –43 dBm will result in third-order intermodulation components that must be less than –113 dBm (referred to the input) so there is still 11 dB of SNR at the input. Thus, the third-order components (at –113 dBm) are 70 dB below the fundamental components (at –43 dBm). Using (3.46) with P_i at –43 dBm and $[P_1 - P_3]$ = 70 dB results in IIP3 of about –8 dBm, and a 1-dB compression point of about –18 dBm at the input. Thus, the single input blocker at –23 dBm is still 5 dB away from the 1-dB compression point. This sounds safe; however, there will now be gain through the LNA and the mixer. The blocker will not be filtered until after the mixer, so one must be careful not to saturate any of the components along this path.

Now after the signal passes through the front end and is downconverted to the IF and passed through the IF filter, the spectrum will be as shown in Figure 5.14. In this case the signal experiences a 20-dB gain, while the two closest blockers experience a net gain of –10 dB and the third blocker experiences a net gain of –30 dB. If no filtering were applied to the system then the IIP3 of the first IF block would need to be roughly –8 dBm + 20 dB = 12 dBm. With filtering, the IM3 products from the two closest blockers must be lower than –93 dBm. Using (3.47) with P_i at –53 dBm

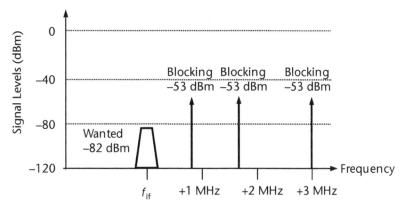

Figure 5.14   Minimum detectable signal and blocker levels after the IF filter.

and $[P_1 - P_3] = 40$ dB results in IIP3 of about –33 dBm and a 1-dB compression point of about –43 dBm for the IF block. Thus, it is easy to see the dramatic reduction in required linearity with the use of filters. Note that the 1-dB compression point is still 10 dB above the level of any of the blocking tones.

5.7   The Effect of Phase Noise and LO Spurs on SNR in a Receiver

The blocking signals can cause problems in a receiver through another mechanism known as reciprocal mixing. For a blocker at an offset of Δf from the desired signal, if the oscillator also has energy at the same offset Δf from the carrier, then the blocking signal will be mixed directly to the IF frequency as illustrated in Figure 5.15.

In general to determine the maximum phase noise allowed at an offset of the adjacent channel one must know the minimum level P_{RFmin} of the desired channel, the maximum adjacent channel power P_{ajdmax}, and the required minimum

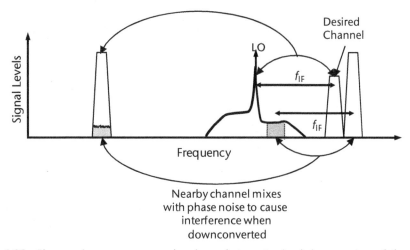

Figure 5.15   Phase noise can cause nearby channels to get mixed down on top of the desired signal.

signal-to-noise ratio SNR_{min} for proper signal detection. After the mixing process, the adjacent channels must have been attenuated by at least

$$RL_{adj} = P_{adj\,max} - (P_{rf\,min} - SNR_{min}) \qquad (5.25)$$

This means that the total power of the LO signal across the bandwidth of the adjacent channel must be lower than the main carrier power by $-RL_{adj}$. Thus, the phase noise across this bandwidth must be less than

$$PN_{max} = RL_{adj} - 10\log(BW)dBc/Hz \qquad (5.26)$$

Alternatively, RL_{adj} directly tells you the maximum level of any discrete spurs over this bandwidth.

Example 5.8:   Calculating Maximum Level of Synthesizer Spurs

For the specifications given in the previous example, calculate the allowable noise in a synthesizer in the presence of the blocking signal in the adjacent channel. Assume that the IF bandwidth is 200 kHz.

Solution:
Any tone in the synthesizer at 800-kHz offset will mix with the blocker which is at –43 dBm and mix it to the IF stage where it will interfere with the wanted signal. The blocker can be mixed with noise anywhere in the 200-kHz bandwidth. Because noise is specified as a density, total noise is obtained by multiplying by the bandwidth, or equivalently in decibels by adding 10 log 200,000 or 53 dB. We note that to be able to detect the wanted signal reliably, as in the previous example, we need the signal to be about 11 dB above the mixed-down blocker. Therefore, the mixed-down blocker must be less than –113 dBm. Therefore, the adjacent channel must be attenuated by: –43 – (–113) = 70 dB thus the maximum synthesizer noise power at 800-kHz offset is calculated as –70 – 53 = –123 dBc/Hz lower than the desired oscillating amplitude.

5.8   DC Offset

DC offset is caused primarily by leakage of the LO into the RF signal path. If the LO leaks into the signal path and is then mixed with itself (called self-mixing), this mixing product will be at zero frequency as shown in Figure 5.16. This will produce a DC signal or DC offset in the baseband that has nothing to do with the information in the modulated signal. This offset will be amplified by any gain stages in the baseband and can saturate the ADC if it is too large. DC offsets can also be caused by PA leakage and jamming signals received from the antenna. This problem is often worse in direct conversion radios where there is usually much more gain in the baseband of the radio and the LO is often at a much higher frequency, thus reducing the LO isolation of the radio.

There are a couple of things that can be done about DC offset. If the radio uses a modulation type where there is not much information at DC (such as an OFDM signal where the first subcarrier does not contain any information), then a blocking

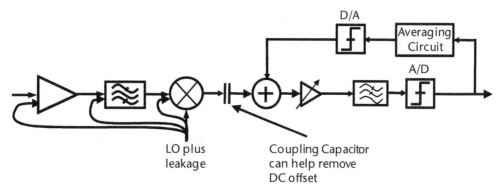

Figure 5.16   LO self-mixing can cause DC offsets.

capacitor can be placed right before the ADC. This will act as a highpass filter and will prevent the DC offset from entering the ADC. However, often this cannot be done in practice because the capacitor will be very large (often prohibitively large), which will also lead to a very slow startup transient event. Because DC offsets are often not time-variant, it may also be possible to calibrate much of it out of the signal path. This can be done by sensing the DC offset in the baseband and adjusting the output from the mixer to compensate for it.

5.9   Second-Order Nonlinearity Issues

Second-order nonlinearity is also very important in transceiver specifications. In this case, the main cause of nonlinearity is the IQ mixer that downconverts the signal to baseband. Consider again the case where there are two in-band interfering signals. If these signals are close to each other in frequency, then the difference between these two frequencies may fall into the baseband bandwidth of the receiver. If the mixer has a finite RF-to-IF isolation, then some faction of this energy may appear at the baseband output and cause interference as shown in Figure 5.17. If this happens, then the signal-to-distortion ratio must still be large enough to make sure that the signal can be detected with sufficient BER. It is usually only the final downconversion stage that is the problem as, prior to the mixer itself, a simple ac coupling capacitor will easily filter out any low-frequency IM2 products produced by the earlier stages in the radio.

It is also important to note that this problem will be more important in direct downconversion radios because in these radios prior to downconversion there cannot be any filtering of in-band interferers, while in a superheterodyne radio in-band interferers can be filtered in the IF stage of the radio making this problem much less severe.

Example 5.9:   IIP2 Calculation

A receiver is trying to detect a signal at 2 GHz with a power level of –80 dBm at the input of the downconversion mixer. An SNR of 15 dB is required and the signal has

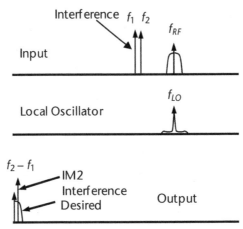

Figure 5.17 Illustration of a second-order product causing distortion.

a bandwidth of 20 MHz. Two interferers are present in band, each with a power level of –20 dBm. They are located at 2,100 MHz and 2,110 MHz. The RF-to-IF isolation of the mixer is 20 dB. Determine the required IP2 for this system.

Solution:
The second-order nonlinearity of the mixer will produce a tone at a frequency of 2,110 MHz – 2,100 MHz = 10 MHz, which will fall into the band of the desired channel. Because the signal strength is –80 dBm, the SNR required is 15 dB and the RF-to-IF isolation is 20 dB, the power of this distortion product must be less than –80 – 15 + 20 = –75 dBm. In this case and using (3.55), we note that $P_2 - P_1 = (-20 - (-75))$ dB = 55 dB. Therefore, the IIP2 must be greater than –20 dBm + 55 dB = 35 dBm.

5.10 Automatic Gain Control Issues

To keep the ADCs simple, it is required to keep their input amplitude relatively constant. Simple ADCs are not equipped to deal with large ranges of input amplitude and therefore it is one of the most important jobs of the radio to provide them with a constant signal amplitude. On the receive side, therefore, as a bare minimum, the radio must provide an AGC range at least equal to the dynamic range of the radio. In addition, radio gain will vary with temperature, voltage supply variations, and process variations. Usually at least an additional 20 dB of gain control is required to overcome these variations. Normally it is possible to use stepped AGC with discrete gain settings (usually spaced by about 3 dB, but maybe less as needed), but some more sophisticated radios may require continuous AGC. It is important to decide where the AGC level will be set and in what order different gain control blocks will be adjusted. At the minimum detectable level, the receiver is set to the maximum gain setting. As the input power starts to rise, it is better to reduce the gain as far towards the back of the radio as possible to have the lowest effect on the noise figure. At high power levels, it is better to adjust the gain at the front of the radio, thus reducing the linearity requirements of parts further down the chain.

On the transmit side, AGC is often simpler than on the receive side. Even simple transmitters must have some AGC to compensate for process, temperature, and supply voltage variations. Typically a simple stepped AGC of about 20 dB is used to make sure that the required power can always be delivered to the antenna. More sophisticated radios will use power control in the transmitter as well to back transmit power off if the receiving radio is close by, thus reducing the required maximum received power on the other side of the link. Usually, a step size of about 4 dB is needed, unless the receive power is used to determine the transmit power, in which case tighter control is needed. The time constant of an AGC loop in a typical gigahertz communication system is usually around 4 to 20 ms.

Usually there are two power detectors employed in AGC as shown in Figure 5.18. One is at the front end of the radio before filtering and it detects broadband power. The second is after the ADC and it detects in-band power. Subtracting the two (after adjusting for the gain) will allow signal power and interferer power to be determined. This indicates how to start to optimize gain control. For instance, if signal power is strong then as a first step, gain can be reduced at the front of the radio first (the LNA if it has a gain step) because SNR will not be the limiting factor. This has the desirable outcome of reducing the linearity stress on components down the chain. If the signal is weak, but there is much interference, then there is no choice but to back off gain at the back end of the radio to keep the SNR high. Thus, in this case, the linearity of the front end will be put to the test. To avoid stability problems, each AGC step should have some hysteresis built into it. This will avoid flipping back and forth due to noise when close to a transition boundary.

5.11 Frequency Planning Issues

With direct downconversion radios, there is little choice in what frequencies will be present in the radio; however, with superheterodyne radios one must choose the intermediate frequencies as well as choosing whether the LOs will be high-side or low-side injected. These frequencies must be carefully chosen, especially in a full-duplex transceiver where transmit and receive tones will be present at the

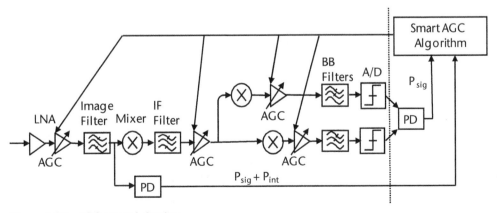

Figure 5.18 AGC control circuitry.

same time. The problem is that often these signals or harmonics of these signals can mix together and if the mixing products fall into the receive band or the transmit band, then this can reduce the SNR of the signal and cause decreased BER performance.

The choice of the first LO is often the first thing to consider. Often in a full-duplex transceiver it is highly desirable to share the LO between the receive path and the transmit path. Thus, in this case once the receive side IF frequency is determined, the transmit side IF will also be determined by default. In a half-duplex transceiver, often the LO will still be shared between the two paths and often out of convenience the IF will be the same in both paths.

To set the LO frequency, we must first determine the receiver IF frequency, which is the more important of the two IF frequencies. This involves a number of trade-offs and considerations. Choosing a low-frequency IF will be good for IF filter design that may otherwise have a high Q requirement. However, a low IF will make the image filter harder to design so some compromise between these two requirements is required. In-band interference between a strong transmit tone and the receive band is also a major consideration. The Tx channels are potentially more troublesome than other random interfering signals because of a limited ability to filter them. After all the Tx band is likely close in frequency to the Rx band and these frequencies have to pass through any filter shared by both the Tx and Rx paths. This problem can be avoided by making the IF less than the band spacing between the TX and RX bands, so

$$f_{IF} < f_{BS} \tag{5.27}$$

If the IF is chosen larger than this value, a nearby TX signal could act as an LO signal and mix an undesired RX channel on top of the desired channel at the IF as shown in Figure 5.19. If there is a need to make the IF higher than the band separation, then the IF should be larger than

$$f_{IF} > 2f_{BW} + f_{BS} \tag{5.28}$$

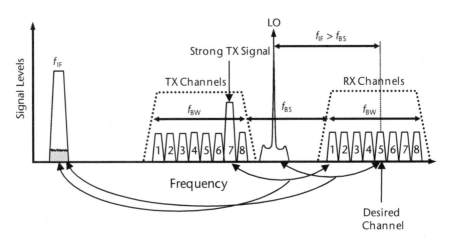

Figure 5.19   Illustration of a bad choice of IF.

This will ensure that even the lowest-frequency TX channel mixing with the highest RX channel will be unable to generate interference at the IF as shown in Figure 5.20.

Another effect that sets a lower limit on the IF is the IF/2 problem. In this case another unwanted channel is mixed down to half the IF. At this point, a second-order nonlinearity can mix this term on top of the IF itself. Because these in-band channels cannot be filtered, it is desirable to set the IF so this cannot happen. To make sure this cannot happen, the IF must satisfy

$$f_{IF} > 2f_{BW} \tag{5.29}$$

This can be illustrated as shown in Figure 5.21.

5.11.1 Dealing with Spurs in Frequency Planning

Interaction of different signals and their harmonics inside a radio is inevitable; however, one thing that the radio system's engineer can do is to mitigate how many potentially harmful unwanted signals fall onto particularly sensitive frequency bands. In general, all signals present and their harmonics must be considered. For RF signals it may be sufficient to consider only the first 8 to 12 harmonics, with IF signals you may have to consider up to 14 to 20 harmonics and with a reference signal (typically 40 MHz or less) 30 to 40 harmonics may be necessary. These signals can all mix with one another. Generally, it is the low-frequency mixing products that must be considered. Thus, for any two fundamental tones in the radio f_a and f_b, we are interested in

$$f_{spur} = mf_a, nf_b, mf_a \pm nf_b \tag{5.30}$$

where m and n are integers.

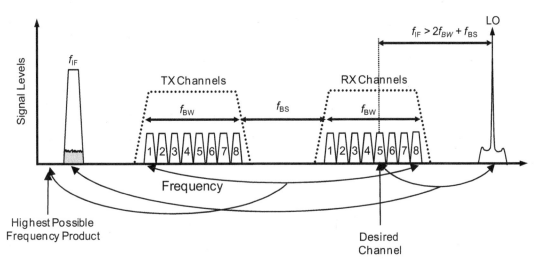

Figure 5.20 A better choice for an IF.

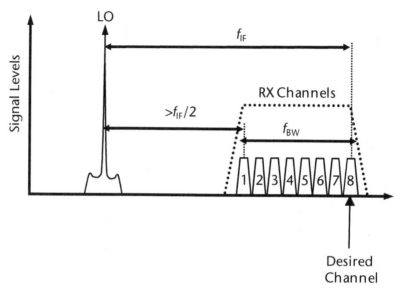

Figure 5.21 Illustration how to avoid the IF/2 problem in a receiver.

Once all the spurs have been located then we need to find out if any fall into:

1. The receive band;
2. The transmit band;
3. The image band;
4. The IF band;
5. The IF/2 band;
6. Close to either the RF or IF LO frequencies.

Any tones falling onto a signal path frequency has the possibility of jamming the radio. We are also concerned with signals that fall onto or close to the LO frequencies. Tones at or near these frequencies could increase phase noise or, worse, pull the LO off frequency.

Example 5.10:   Frequency Planning a Full-Duplex Superheterodyne Radio

Design a full-duplex radio to operate at 7 GHz. It will contain 11 channels. The channels will be 20 MHz wide. The Tx band will be from 6.8 GHz to 7 GHz and the Rx band will be from 7.2 GHz to 7.4 GHz. Assume that you want a high performance (superheterodyne radio) and that your crystal oscillator (XTAL) reference is at 40 MHz. Choose a frequency plan for the radio. For frequency planning, consider three harmonics of the transmitted signal, RFLO, and received signal, seven harmonics of any IF signals, and 15 harmonics of the XTAL.

Solution:
First, an LO must be chosen. Assuming that the LO will be shared between the Tx and Rx bands, we need to decide first on high-side or low-side injection. We choose high-side injection arbitrarily. Now assuming that a very low IF is not an option due to the image filter, the minimum IF from (5.28) we could choose would be:

$$f_{IF} > 2f_{BW} + f_{BS} = 2(200 \text{ MHz}) + 200 \text{ MHz} = 600 \text{ MHz}$$

We will choose this IF (600 MHz) for the Rx. That sets the LO to operate between 7.8 and 8.0 GHz. This also sets the Tx IF at 1 GHz. Now we need to exhaustively look at all harmonics and their mixing products in this radio. Table 5.2 lists signals present in the radio.

Now we need to look at the mixing products of all these frequencies and their harmonics and determine if any of them may be a problem at a critical frequency. This will involve a really large number of calculations and plots. To perform this, some programing language or a spread sheet should be employed. After all these calculations were performed, the results were plotted and all but the potentially troublesome mixing products and harmonics were removed from the plot make it more readable. The receive band spur frequencies are plotted in Figure 5.22, and the transmit band spur frequencies, image band spur frequencies, IF band spur frequencies, and RFLO band spur frequencies are plotted Figures 5.23 through 5.26. In these figures the horizontal axis represents the channel to which the radio is currently tuned. The lines are plotted to show the movement of a spur's frequency as the channel is changed. The labels use shorthand to list either the source of the spur and its harmonic or the two sources of the spur and their harmonics. For instance, Rx1 is the first harmonic of the RF signal and TxIFXTALSum 7,6 is the seventh harmonic of the transmitter IF signal mixed with the sixth harmonic of the crystal oscillator to produce a spur at the sum (rather than the difference) of these two frequencies. Note that TxIF is the IF of the transmitter, RxIF is the IF of the receiver, XTAL is the crystal oscillator, Tx is the RF transmit frequency, and Rx is the RF receive frequency.

Looking more closely at the receive band spurs in Figure 5.22, we note that the line labeled Rx1 represents the desired receive signal. We have two other lines representing spurs that directly fall on top of the Rx1 line; these are TxXTALSum1,10, which is the transmit component at a 400-MHz offset from the receiver, mixing with the tenth harmonic of the crystal; hence, the product is at 400 MHz, resulting in a spur directly on top of the receive signal. The other interfering spur is RFLOXTALDif1,15. In this case, the LO tracks the receive signal, but is above it by 600 MHz. This is mixed with the fifteenth harmonic of the crystal frequency, which is at 600 MHz, and the difference is directly on top of the receive signal. All other spurs shown are at fixed frequencies so they will each fall on top of only one particular channel.

In Figure 5.23 for the transmit band, we note that the line labeled Tx1 is the desire transmit signal. Similar to the Rx band, a spur falls directly on top of the Tx signal from the Rx signal at 600 MHz away mixed with the tenth harmonic of the

Table 5.2 Summary Signal Tones Present in the Radio

Signal Source	Ch 1	Ch 2	Ch 3	Ch 4	Ch 5	Ch 6	Ch 7	Ch 8	Ch 9	Ch 10	Ch 11
Receiver signal	7,200	7,220	7,240	7,260	7,280	7,300	7,320	7,340	7,360	7,380	7,400
Transmitter	6,800	6,820	6,840	6,860	6,880	6,900	6,920	6,940	6,960	6,980	7,000
RFLO	7,800	7,820	7,840	7,860	7,880	7,900	7,920	7,940	7,960	7,980	8,000
TxIF	1,000	1,000	1,000	1,000	1,000	1,000	1,000	1,000	1,000	1,000	1,000
RxIF	600	600	600	600	600	600	600	600	600	600	600
XTAL	40	40	40	40	40	40	40	40	40	40	40

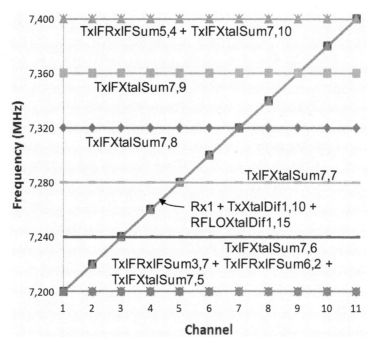

Figure 5.22 Plot of spurs that appear in the receiver band. The labels are a shorthand indicating the source or sources of the spur and which harmonic where Rx1 is the desired fundamental received signal. Tx is the RF transmit frequency, Rx is the RF receive frequency, XTAL is the crystal oscillator, TxIF is the IF of the transmitter, and RxIF is the IF of the receiver. Note there is at least one interfering signal that falls directly on top of each desired frequency.

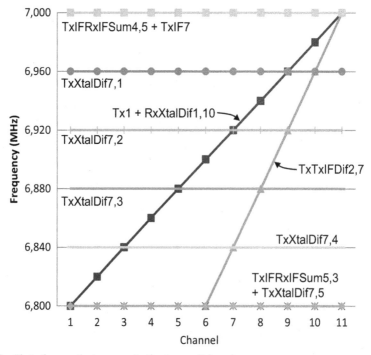

Figure 5.23 Plot of spurs that appear in the transmit band.

crystal. All other spurs shown each will fall on top of the transmit signal exactly in one particular channel.

In the image band, shown in Figure 5.24, the image band extends from 8,400 through 8,600 MHz, which is the receive frequency plus twice the receive IF. Spurs that fall directly on top of the image signal are a result of the second harmonic of the LO mixing with the receive signal, and the RF LO mixing with the fifteenth harmonic of the crystal frequency. All other spurs shown each will fall on top of the image frequency exactly in one particular channel.

Figure 5.25 shows the spurs in the IF bands. For the Tx IF frequency of 1,000 MHz, there are five other sources of spurs, while for the Rx IF frequency of 600 MHz there are six other sources of spurs. In addition, in the IF/2 band at 300 MHz, there is interference in channel 6 from mixing between the receive frequency at 7,300 MHz with the seventh harmonic of the transmit IF at 7,000 MHz.

Figure 5.26 shows the spurs in the RxLO frequency band from 7,800 MHz for channel 1 to 8,000 MHz for channel 11. It can be seen that there are three directly interfering sets of spurs. Two of these result from mixing of the receive frequency with the receive IF signal or with the fifteenth harmonic of the crystal frequency. The third is due to the transmit frequencies mixing with the transmit IF frequency. Other sources of spurs are the mixing of transmit IF with receive IF frequencies, one interfering with channel 1 and the other interfering with channel 11.

It can be noted that many of the spurs are due to interaction with the TxIF as well as interaction with the XTAL. Therefore, after looking carefully at the mixing products in all these figures, it might be a much better choice to low-side inject the RFLO rather than high-side inject it. As well, increasing the RxIF to 620 MHz will move it away from some of the XTAL spurs. After doing both of these things, the fundamental tones are now summarized in Table 5.3.

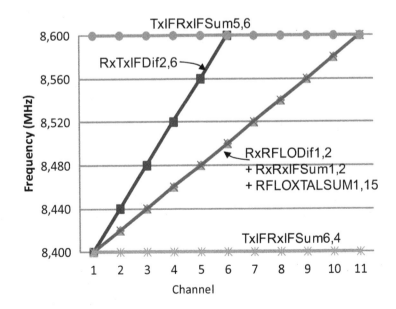

Figure 5.24   Plot of spurs that appear in the image band.

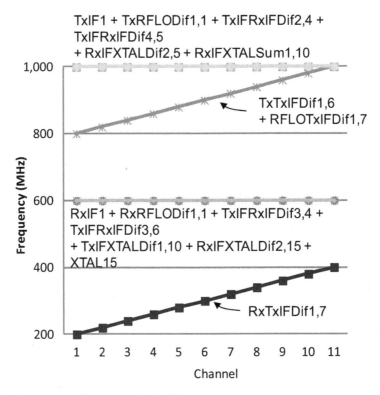

Figure 5.25 Plot of spurs that appear in the IF bands.

Now the spurs are analyzed once more and the results are presented in Figure 5.27, Figure 5.28, and Figure 5.29. Note that because the receive and transmit band are separated by 400 MHz, mixing with the tenth harmonic of the crystal oscillator frequency will result in spurs in the other band, but these are the only spurs seen in these bands. Therefore, this is a much better choice of frequency plan than the

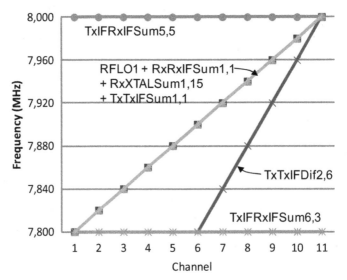

Figure 5.26 Plot of spurs that appear in the RFLO bands.

Table 5.3 Summary Signal Tones Present in the Radio with a Low Side Injected RFLO

Signal Source	Ch 1	Ch 2	Ch 3	Ch 4	Ch 5	Ch 6	Ch 7	Ch 8	Ch 9	Ch 10	Ch 11
Receiver Signal	7,200	7,220	7,240	7,260	7,280	7,300	7,320	7,340	7,360	7,380	7,400
Transmitter	6,800	6,820	6,840	6,860	6,880	6,900	6,920	6,940	6,960	6,980	7,000
RFLO	6,580	6,600	6,620	6,640	6,660	6,680	6,700	6,720	6,740	6,760	6,780
TxIF	220	220	220	220	220	220	220	220	220	220	220
RxIF	620	620	620	620	620	620	620	620	620	620	620
XTAL	40	40	40	40	40	40	40	40	40	40	40

first try; however, much more work could be done to further improve the frequency plan. This would likely involve many more iterations where frequency choices were changed or refined and evaluating what spurs remained to be dealt with.

The IF bands and IF/2 bands are also now much cleaner.

5.12 EVM in Transmitters Including Phase Noise, Linearity, IQ Mismatch, EVM with OFDM Waveforms, and Nonlinearity

Error vector magnitude (EVM) is a very important way to measure how accurately a transmitter has reproduced the vectors that correspond to the data being transmitted. EVM is another way to measure the signal to noise and distortion ratio of the signal. The term EVM is used more often in transmitters compared to receivers as a measure of modulation accuracy instead of SNR.

While EVM is more typically measured in a transmitter, many of the concepts discussed in this section are also applicable to receivers as well. When a signal is

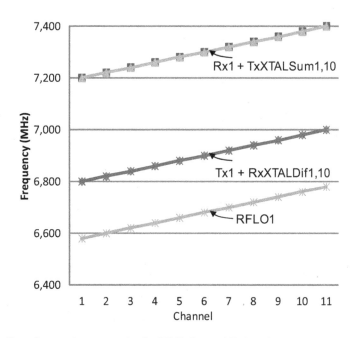

Figure 5.27 Plot of spurs that appear in the RFLO, Rx, and Tx bands.

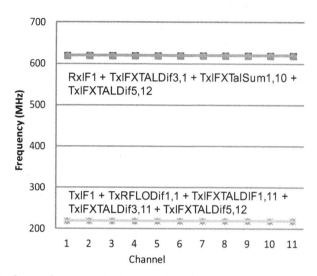

Figure 5.28 Plot of spurs that appear in the IF and IF/2 bands.

generated in the BBSP of the transmitter, it is nearly perfect. Once the signal passes through the transmitter, due to a large number of imperfections the waveform will be distorted and contain the original information plus some distortion and will therefore have a finite signal-to-distortion level (SDR). SDR is usually measured as EVM (discussed shortly). If the SDR is not larger than the required SNR at the receiver input, then information transmission will not be successful even over zero distance. A good rule of thumb may be that the SDR should be at least 6 dB higher than the required SNR. As the signal is transmitted over distance, it is exposed to a noisy channel. The signal power drops and while the SDR stays constant, the SNR gets reduced. At the maximum transmit distance, it is expected that the SNR may be less than the SDR, but it should still be greater than the minimum SNR required for detection. The signal then passes through the receiver where the signal power is

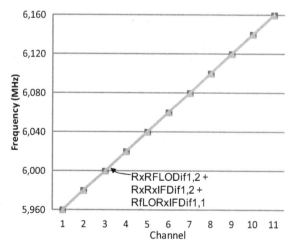

Figure 5.29 Plot of spurs that appear in the image band.

again amplified, but the SNR and SDR are both reduced once more due to noise and nonidealities in the receiver. This process is illustrated in Figure 5.30.

In any system with limited performance, some errors will always be introduced and the vector that is actually transmitted will be different by some amount (an error) from what was intended as shown in Figure 5.31. The instantaneous value of the EVM is defined as the ratio of the magnitude of the reference vector to the magnitude of the error vector

$$EVM_i = \frac{|e_i|}{|a_i|}$$ (5.31)

where e_i is the ith error vector and a_i is the ith reference vector. Normally EVM is averaged over a large number of data samples N to come up with an average or nominal value:

$$EVM = \sqrt{\frac{1}{N}\sum_{i=1}^{i=N}\left(\frac{|e_i|}{|a_i|}\right)^2}$$ (5.32)

There are many sources of EVM in a transmitter and the overall effect of various sources can be added together as

$$EVM_{tot} = \sqrt{(EVM_1)^2 + (EVM_2)^2 \ldots + (EVM_M)^2}$$ (5.33)

Example 5.11:   Determining the Required EVM for a Transmitter

A transmitter is using an OFDM, 16QAM modulation with 64 subcarriers. The required BER for transmission is 10^{-3}. What is the required EVM of the transmitter? Assume a 6-dB safety margin.

Solution:
For a BER of 10^{-3} with 16QAM a SNR of 17 dB is required. With a 6 dB safety margin that means an EVM of 23 dB or about 7% is required. Remember that EVM is basically just a different name for SNR in the transmitter, because signal corruption in a transmitter comes more often from linearity and mixing than noise.

One source of EVM is synthesizer phase noise. As discussed in Chapter 3, the phase noise can be integrated to give an rms value for the phase variation in radians ($IntPN_{rms}$). This phase variation will affect the angle of the reference vector

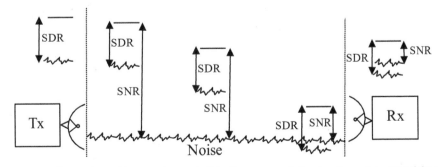

Figure 5.30 Illustration of how noise and distortion creeps into signals as they travel from the source to the destination.

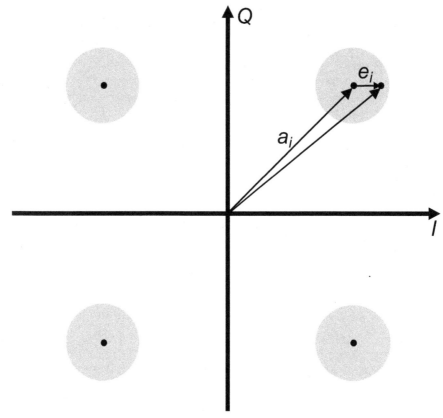

Figure 5.31 Illustration of EVM using QPSK as an example modulation.

and will move it away from its ideal value. Thus, in this case the phase noise will generate an error vector that is orthogonal to the reference vector as shown in Figure 5.32. In this case the error vector will have a magnitude of

$$|e_{iPN}| = a_i \sin\left(IntPN_{rms}\right) \approx a_i IntPN_{rms} \tag{5.34}$$

Thus the EVM due to phase noise is

$$EVM_{PN} = \frac{|a_i| \cdot IntPN_{rms}}{|a_i|} = IntPN_{rms} \tag{5.35}$$

Example 5.12: Determining Phase Noise for Required EVM

What phase noise is required to achieve an EVM of 1%? Assume that the baseband signal bandwidth is 500 kHz and that the phase noise of the synthesizer is flat across the entire bandwidth.

Solution:
An EVM of 1% or 0.01 means that the integrated rms phase noise must be 0.01 rads using (5.35). If the phase noise is flat across the entire band of 500 kHz, that means that the phase noise must be 20 nrads/Hz or –77 dBc/Hz.

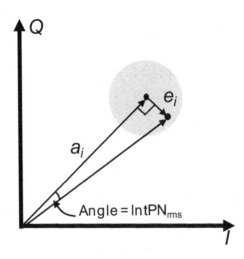

Figure 5.32 Phase noise can cause EVM by rotating the vector.

Note that a phase mismatch between the I and Q paths either created by the synthesizer or a mismatch in the baseband paths will behave exactly the same way in regards to EVM as phase noise does. In this case if there is a phase mismatch in the LO of θ_{LO}, then the EVM due to IQ phase mismatch will be

$$EVM_{LO} = \theta_{LO} \tag{5.36}$$

Another source of EVM is gain mismatch in the baseband I and Q paths leading up to the mixers. If there is some random shift in amplitude from the ideal amplitude in the I and Q paths δ such that the magnitude of the I path becomes $I(1 + \delta)$ and the magnitude of the Q path becomes $Q(1 - \delta)$, then the error vector will have a magnitude of

$$e_i = \sqrt{(\delta I)^2 + (\delta Q)^2} \tag{5.37}$$

The reference vector will have a magnitude of

$$a_i = \sqrt{I^2 + Q^2} \tag{5.38}$$

In this case the EVM will be simply

$$EVM_{IQ} = \delta \tag{5.39}$$

Example 5.13: EVM Simulation of a 16QAM Signal

Simulate the EVM of a 16QAM signal experiencing a phase mismatch of 3° and an IQ amplitude imbalance of 10%. Compare to the simple theory presented.

Solution:
For this simple simulation an ideal up converter is used to generate an RF signal from IQ baseband signals. No filtering or bandwidth limitations are considered. The only nonidealities are the amplitude and phase mismatch. Thus, in the simulation

the I channel's amplitude is increased by a factor of 1.1 and the Q channels amplitude is decreased by a factor of 0.9. A phase error of 3° is 0.0524 radians. The otherwise ideal LOs are now generated as $\cos(\omega_{LO}t - 0.0524)$ and $\sin(\omega_{LO}t + 0.0524)$ to model the phase error. Thus the EVM from this simulation is expected to be:

$$EVM_{LO} = 10\log\left(\theta_{LO}^2 + \delta^2\right) = 10\log\left(0.0524^2 + 0.1^2\right) = -18.9 \text{ dB}$$

This simulation was run with the aid of MATLAB. The resulting EVM plot is shown in Figure 5.33. The simulated EVM was –19.8, which is 0.9 dB better than that predicted by the simple theory above.

EVM can also be caused by carrier feedthrough. Carrier feedthrough can be due to finite isolation from LO to RF and from LO to IF. Whatever the source, any energy from the carrier will distort the waveform and cause EVM. If we define carrier suppression C_s as the ratio of the desired RF power P_t to the LO leakage P_{CFT}, then the EVM due to carrier feedthrough is given by

$$EVM_{CFT} = \sqrt{\frac{P_{CFT}}{P_t}} = \sqrt{C_s} \tag{5.40}$$

Note that P_{CFT} will be a vector with a random phase relative to the reference vector, but only the magnitude is important.

Linearity in the transmitter may also be a consideration in regards to EVM, depending on the type of modulation used. If the system uses a phase-only modulation,

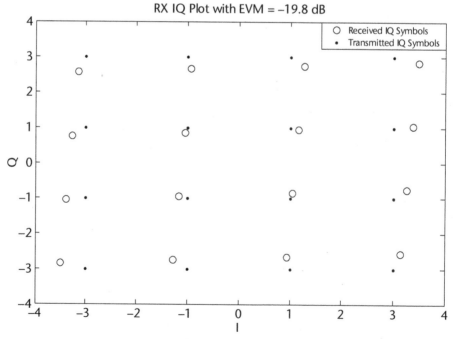

Figure 5.33 Constellation plot of a 16QAM waveform after experiencing IQ amplitude and phase mismatch.

linearity is of much less concern. With amplitude sensitive modulations like QAM it is best to operate the output with a power level that is well backed off from the 1-dB compression point of the transmitter. If it is assumed that the transmitter is reasonably linear and that the linearity can be modeled by a third-order power series as shown in Chapter 3, then the instantaneous EVM would be given by

$$EVM_i = \frac{\frac{3}{4}k_3 v_{RFi}^3}{k_1 v_{RFi}} = \frac{0.285 \frac{v_{RFi}^2}{v_{1dB}}}{Gain} \tag{5.41}$$

where v_{RFi} is the instantaneous RF voltage amplitude at the input of the amplifier. EVM_i must be computed for each constellation point and then divided by the total number of constellation points

$$EVM_{nonlin} = \frac{1}{n}\sum_{i=1}^{n} \frac{0.285 \frac{v_{RFi}^2}{v_{1dB}}}{Gain} \tag{5.42}$$

Example 5.14:   Determining the EVM of a 16QAM Transmitter After Passing Through the Power Amplifier

A transmitter is used to transmit a 16QAM signal. The constellation in one quadrant is: (1,1), (3,1), (1,3), and (3,3). Choose a 1-dB compression point for the RF PA and calculate the EVM and then compare this to simulation results.

Solution:
With this modulation scheme, we need to first determine the signal amplitudes. With a random bit stream it can be expected that symbols at a distance of (1,1) or a voltage amplitude of 1.414V will be transmitted 25% of the time, symbols at an amplitude of 3.162V using points (3,1) or (1,3) will be transmitted 50% of the time, and that symbols at an amplitude of 4.243V will be transmitted 25% of the time.

   A 1-dB compression point needs to be selected. As a starting point, 4.5V is selected. Also assume that the amplifier gain is 1. This should provide an EVM of

$$EVM_{nonlin} = \frac{1}{4}\left(0.285 \cdot \frac{1.414^2}{4.5} + 2 \cdot 0.285 \frac{3.162^2}{4.5} + 0.285 \frac{4.243^2}{4.5}\right) = 14.1\%$$

This was simulated with no band-limiting filters and after ideal demodulation the output constellation for 1,000 symbols transmitted is shown in Figure 5.34. Note that the outer points are far more compressed than the inner ones. The actual simulated EVM was 16.4%, which is close to the theoretical value predicted above. Note that this is still a large number and could be further improved by increasing the linearity of the transmitter.

   Systems that use OFDM have a different set of linearity requirements in regards to EVM than systems that use simpler modulation schemes. If we assume that the transmitter is operating well below the 1-dB compression point and the linearity can be well described by third-order products, the dominant contribution to EVM degradation due to linearity of the transmitter will be triple-order beats of

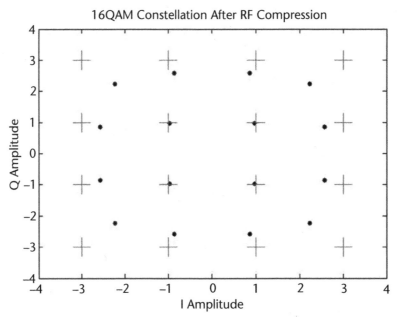

Figure 5.34 Constellation plot of a 16QAM modulated waveform after passing through a nonlinear amplifier.

all subcarriers dumping energy on top of the subcarrier of interest. The power in a triple-beat tone as shown in Chapter 3 is given by

$$TB = OIP_3 - 3(OIP_3 - P_s) + 6 = 3P_s - 2OIP_3 + 6 \tag{5.43}$$

where P_s is the power of the tones generating the triple beat (in this case the power in one subcarrier) and OIP_3 is the output referred third-order intercept point, both given in dBm. The EVM can then be computed as the ratio of the power of the subcarrier to the power of all triple beats dumping power into that channel. The difference between the power of the triple beat tones and the desired channel power is therefore

$$TB - P_s = 2P_s - 2OIP_3 + 6 + 10\log\left(\frac{3}{8}N^2\right) \tag{5.44}$$

Thus, the EVM due to transmitter nonlinearity is given as

$$EVM_{lin} = 10^{\frac{\left[2P_s - 2OIP_3 + 6 + 10\log\left(\frac{3}{8}N^2\right)\right]}{20}} \tag{5.45}$$

where N is the number of subcarriers.

Note that in a multicarrier system we have N subcarriers. Thus, the total transmitted power is related to P_s by

$$P_t = P_s + 10\log N \tag{5.46}$$

Thus, (5.45) can be rewritten in terms of transmit power as

$$EVM_{lin} = 10^{\frac{\left[2P_t - 10\log N^2 - 2OIP_3 + 6 + 10\log\left(\frac{3}{8}N^2\right)\right]}{20}}$$

$$= 10^{\frac{\left[2P_{tx} - 2OIP_3 + 6 + 10\log\left(\frac{3}{8}\right)\right]}{20}} \tag{5.47}$$

Getting rid of the exponent this can be written more simply as

$$20\log(EVM_{lin}) = 2P_t - 2OIP_3 + 6 + 10\log\left(\frac{3}{8}\right) \tag{5.48}$$

Equation (5.48) has been verified through extensive simulations and comparison to measured results by the authors. However, note that if the output power is close to the 1-dB compression point, this formula will become less accurate. Note also that narrowband systems rely on filtering to minimize the number of tones that can intermodulate making TB tones much less of an issue. However, OFDM uses multiple subcarriers that are always present; hence, it is much more like a video system in which a whole range of tones can intermodulate, even though EVM or BER and not CTB is typically used as a measure of performance.

Example 5.15:   Determining Transmitter Linearity

A transmitter is required to transmit at a power level of 25 dBm and is using a 64QAM OFDM modulation with 64 subcarriers. What is the required linearity of the transmitter to achieve an EVM of 2.5%?

Solution:
Here we can make use of (5.47):

$$-32 = \left[2(25) - 2OIP_3 + 6 + 10\log\left(\frac{3}{8}\right)\right]$$

$$OIP_3 = 41.9 \text{ dBm}$$

Therefore, the output 1-dB compression point would be about 32.3 dBm, which gives 7.3 dB of backoff.

If the transmitter is backed off and transmitting at a low-power level, then the noise in the transmitter circuits can become an important contribution to the EVM of the system. If the total noise power density coming from the transmitter is N_0, then the EVM contribution from noise is given by

$$EVM_{noise} = \sqrt{\frac{N_0 \cdot BW}{P_t}} \tag{5.49}$$

where BW is the bandwidth of the channel.

Filters in the signal path can also be a source of EVM if their bandwidth is close to the signal bandwidth. Thus, it is usually only necessary to consider baseband filters or perhaps an IF filter if it is only a couple of channels wide. As it is impossible to build an ideal filter, any real filter will have a nonzero impulse response during adjacent bit periods. Thus, power will be spread from one symbol period to another. This is called intersymbol interference (ISI) and can also lead to EVM. Knowing a filter's impulse response $h_f(t)$, its contribution to EVM can be found as

$$EVM_{filter} = \sqrt{\frac{\sum\limits_{k=-\infty,k\neq0}^{k=\infty} \left|h_f\left(t_o + kT_s\right)\right|^2}{\left|h_f(t_o)\right|^2}} \qquad (5.50)$$

where T_s is the sampling period. Note that in practice a good approximation may be possible taking into account only a few adjacent bit periods.

5.13 Adjacent Channel Power

In any transmitter there can be power transmitted beyond the bandwidth of the channel. This is an undesired effect as it has the potential to interfere with other communications. Thus, there are stringent requirements on how much power leakage can take place in any transmitter. The adjacent channel power ratio (ACPR) is defined as

$$ACPR = \frac{\int\limits_{f_1}^{f_2} PSD(f)df}{\int\limits_{f_{cb}-\frac{BW}{2}}^{f_{cb}+\frac{BW}{2}} PSD(f)df} \qquad (5.51)$$

where BW is the bandwidth of the channel. Note that the bandwidth over which the adjacent power is measured is not always equal to the channel bandwidth and is often some faction of the complete channel as shown in Figure 5.35.

While it would be possible to have a low ACPR with a single carrier modulation, this would probably be due to the transmitter being pushed beyond the 1-dB

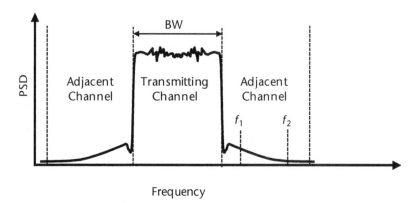

Figure 5.35 Figure showing adjacent power leakage.

compression point. Simple analytical theory would not be a very accurate tool at this stage and simulation would be needed to determine the actual value of ACPR. With multicarrier signals ACPR becomes much more of an issue even if the transmitter is fairly linear compared to the signal power levels.

To start, consider the case where two carriers are present in the channel. They are spaced so that the IM3 products will fall in the channel above and below the transmitting channel. These IM3 tones will result in a finite ACPR value. Assuming that the power level of each tone is P_t and that the OIP_3 of the transmitter is known, the power of the IM3 tones can be calculated as

$$IM3 = OIP_3 - 3(OIP_3 - P_t) \tag{5.52}$$

Thus, in this case because the total transmit power will be $2P_t$, the ACPR is given by

$$\begin{aligned} ACPR &= OIP_3 - 3(OIP_3 - P_t) - 2P_t \\ &= P_t - 2OIP_3 \end{aligned} \tag{5.53}$$

As each tone represents half the total transmit power P_{TR}, this can also be written as

$$ACPR = \frac{P_{TR}}{2} - 2OIP_3 \tag{5.54}$$

Even though this result is derived for two tones, in general, with N tones this result will still hold. Each tone gets $1/N$ of the total power and each will generate a smaller IM3 tone, but the total will add up to be almost the same in the end. In this case the result will be more like noise and will be distributed across the adjacent channel. If only a fraction of the adjacent channel is used, then the ACPR can be given as

$$ACPR \approx \frac{P_{TR}}{2} - 2OIP_3 + 10\log\left(\frac{f_2 - f_1}{BW}\right) \tag{5.55}$$

Example 5.16:   ACPR Requirements

A radio is required to receive a signal with a SNR of 10 dB. If the minimum detectable signal is –80 dBm and the maximum adjacent power is –40 dBm, what is the maximum allowable ACPR for the radio to operate properly?

Solution:
If the minimum detectible signal is –80 dBm and requires a SNR of 10 dB, the maximum interference power is –90 dBm. If the adjacent signal power is –40 dBm, the ACPR of that signal must therefore be –40 dBm – (–90 dBm) = 50 dBc.

Example 5.17:   Determining ACPR

Consider a radio using a 32-carrier OFDM signal where at baseband each subcarrier uses a center frequency of 1, 2, 3, …, 32 Hz. The subcarriers are modulated with simple BPSK and are then upconverted to a 200-Hz transmit frequency. If the OIP3 of the transmitter is 27.8 dBV and the power density in the channel is –5 dBV/Hz, what is the ACPR if the adjacent power is measured in a 10-Hz BW? Compare to simulation.

Solution:
With 32 subcarriers and a maximum carrier frequency of 32 Hz, we can estimate the passband bandwidth at 34 Hz assuming that each subcarrier has a bandwidth of about 1 Hz and allowing for some guard bands. Thus, the total transmit power will be

$$P_{TR} = -5dBV/Hz + 10\log(34) = 10.3 \text{ dBV}$$

Now the ACPR can be computed:

$$ACPR \approx \frac{10.3}{2} - 2(27.8) + 10\log\left(\frac{10}{34}\right) = -55.8 \text{ dBc}$$

Now this waveform must be created and simulated. The baseband waveform is ideally band-limited and modulated to the desired carrier frequency. The frequency spectrum is shown in Figure 5.36.

Now this waveform is passed through a nonlinear circuit with an OIP3 of 27.8 dBV and the resulting waveform is shown in Figure 5.37. Note the growth of

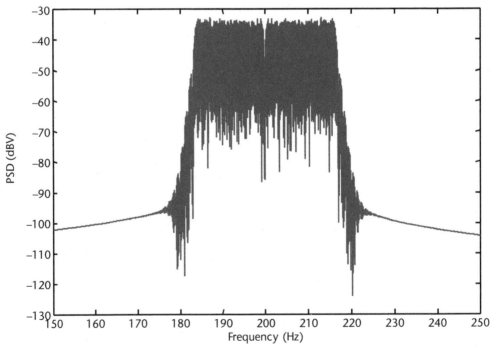

Figure 5.36 Frequency spectrum of a 32-subcarrier band-limited passband OFDM signal. The resolution bandwidth of this plot is 0.001 Hz.

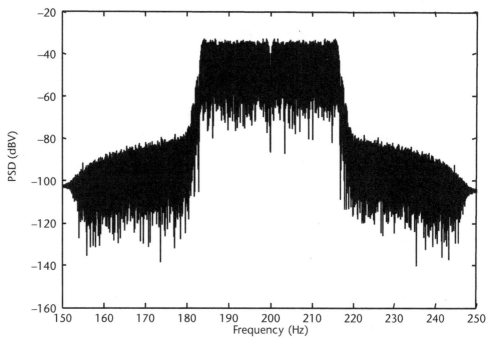

Figure 5.37   Frequency spectrum of a 32-subcarrier bandlimited passband OFDM signal after it has been passed through a nonlinear amplifier. The resolution bandwidth of this plot is 0.001 Hz.

sidebands in the adjacent channels. This nonlinearity is also only modeled up to the third order so it is simpler than a real device. Note that the resolution on these two plots is 0.001 Hz. Therefore, looking at the adjacent channel between 165 and 175 Hz, the power level is about –85 dBV/0.001 Hz or –55 dBV/Hz or –45 dBV over 10 Hz of BW. The channel power is 10 dBV so the ACPR is –55 dBc, which is very close to the predicted value given above.

5.14   Important Considerations in Analog-to-Digital Converters (ADC) and Digital-to-Analog Converters (DAC)

When providing specifications for an ADC, a system designer needs to specify the number of bits, maximum input voltage swing, the required jitter on the clock driving the ADC, and the sampling frequency. To determine what performance level is required by the ADC, a number of issues should be addressed. First, the AGC step size will affect ADC performance requirements. The larger the step size, the larger the input signal amplitude variation. If the AGC does not cover the whole dynamic range of the radio, then the ADC will have to handle larger signals at the maximum input level and smaller signals at the minimum input level. Obviously the SNR of the ADC must be at least as high as the minimum required SNR of the signal. If the signal has a nonunity peak-to-average ratio, then the ADC must be designed to handle the maximum signal level without introducing distortion. There may also be a DC offset in the baseband of the radio. If it is present then this will be an added signal that must be accommodated without distorting or saturating the ADC. The quality of the baseband filters will also play a role in determining the requirements of the ADC. If out-of-band interferers are not completely filtered by the BB filters, then the ADC will have to handle these signals as well without saturating or aliasing them into the receive band.

DACs are similar to ADCs; however, they deal with known signals while ADCs have to be able to deal with less ideal inputs. However, DACs have to be concerned with a number of issues as well. First, harmonics of the sampling clock must be considered as these may cause adjacent channel power. The quality of the transmit baseband filters can be traded off against the amount of aliasing to higher frequencies that can be allowed in the DAC. The dynamic range of the DAC must be large enough to handle the peak-to-average ratio of the signal as well as still have a high enough SNR so that the quantization noise of the DAC does not become the limiting factor in determining the EVM of the transmitter. In the next section more details of DACs and ADCs will be given to better understand how to specify their performance levels.

5.15   ADC and DAC Basics

While it is not the goal of this book to give design details of building blocks, enough detail will be given here to better understand how to compute the required performance levels of ADCs and DACs. A basic ADC consists of a quantizer and a sample-and-hold circuit as shown in Figure 5.38. While many ADCs may be much more complex, this will serve to illustrate many key basic points.

A quantizer converts a continuous analog signal x to a discrete signal according to the quantization rule shown in Figure 5.39, where x is the analog input and y is the quantized output. Note that y has not yet been sampled and therefore is still a continuous signal. The output y is a function of the input, but it has discrete levels at equally spaced intervals Δ. Thus, unless the input happens to be an integer multiple of the quantizer resolution (step size) Δ, there will be an error in representing the input. This error e will be bounded over one quantizer level by a value of

$$-\frac{\Delta}{2} \leq e \leq \frac{\Delta}{2} \tag{5.56}$$

Thus, the quantized signal y can be represented by a linear function with an error e as

$$y = \Delta \cdot x + e \tag{5.57}$$

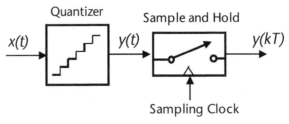

Figure 5.38   A basic block diagram for an ADC.

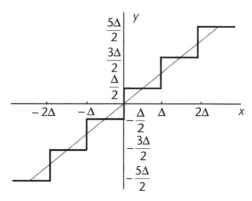

Figure 5.39   Transfer characteristic of a multibit quantizer.

where the step size Δ corresponds to the slope of the straight line shown in Figure 5.39. The quantization error as a function of the input is given in Figure 5.40. Note that the error is a straight line with a slope of $-\Delta$. If the input is "random" in nature, then the instantaneous error will also be random. The error is thus uncorrelated from sample to sample and can hence be treated as "noise". Quantization and the resultant quantization noise can be modeled as a linear circuit including an additive quantization error source as shown in Figure 5.41.

Thus, the quantization noise for a random signal can be treated as additive white noise having a value anywhere in the range from $-\Delta/2$ to $\Delta/2$. The quantization noise has a uniform probability density of

$$p(e) = \begin{cases} \dfrac{1}{\Delta} & \text{if} \quad -\dfrac{\Delta}{2} \le e \le \dfrac{\Delta}{2} \\ 0 & \text{otherwise} \end{cases} \tag{5.58}$$

where the normalization factor $1/\Delta$ is needed to guarantee that:

$$\int_{-\Delta/2}^{\Delta/2} p(e)de = 1 \tag{5.59}$$

The mean square rms error voltage e_{rms} can be found by integrating the square of the error voltage and dividing by the quantization step size,

$$e_{rms}^2 = \int_{-\infty}^{+\infty} p(e)e^2 de = \frac{1}{\Delta} \int_{-\Delta/2}^{+\Delta/2} e^2 de = \frac{\Delta^2}{12} \tag{5.60}$$

Next the sample-and-hold block in Figure 5.38 will be explored. Sampling is the act of measuring a system's output at periodic points in time. Thus, only the

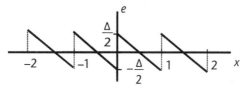

Figure 5.40   The quantization error as a function of the input.

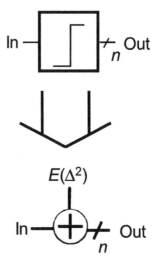

Figure 5.41 Modeling of the quantization as a linear circuit with an additive noise.

system's outputs at the instances when sampling occurs are of interest, and anything the system does in between sampling instances is lost. Sampling is done once per clock cycle, after which the sampled value is held until the next sampling instance. To have a good digital representation of the waveform, it must be sampled fast enough. In fact, the waveform must be sampled at a rate that is at least twice as fast as the highest frequency of interest f_{max}.

$$f_s \geq 2f_{max} \tag{5.61}$$

This is known as the Nyquist sampling rate and although this is the theoretical minimum in practice often choosing f_s to be at least 2.2 times f_{max} is advisable. A sampler also causes delay and therefore it can also be thought of as a delay block. An example of a delay block is a D-flip-flop where the input signal is read on the edge of the clock, while at the same time, the signal stored on the flip-flop (the previous input) is passed on to the next circuit.

We will now describe the process of sampling in more mathematical terms. Consider the continuous time waveform in Figure 5.42. Sampling can be described mathematically as multiplying the waveform by a series of delta functions, sometimes called an impulse train, as shown in Figure 5.42. The waveform $f(t)$ sampled every T time units can be written as:

$$f_s(t) = f(t) \sum_{k=0}^{\infty} \delta(t - kT) = \sum_{k=0}^{\infty} f(kT)\delta(t - kT) \tag{5.62}$$

We note here, without proof, that the Fourier transform of a series of equally spaced delta functions in the time domain (to represent sampling as in Figure 5.42) is a set of delta functions in the frequency domain as shown in Figure 5.43. These are now convolved with the spectrum of the input waveform.

The effect of convolving the input waveform with a delta function is to recenter the waveform at the frequency of the delta function. These impulses can be seen

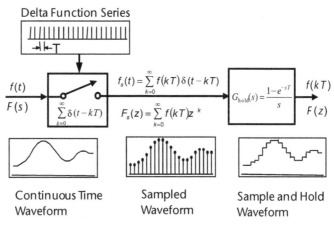

Figure 5.42   Illustration of sampling a waveform.

to occur at dc, at the sampling frequency f_s, at $2f_s$, $3f_s$, and so forth. Thus, the frequency spectrum of the continuous waveform repeats around f_s and around all multiples of the sampling frequency as shown in Figure 5.44. These multiple identical spectral components are the result of mixing of the input with the sampling frequency. This mixing property of sampled systems is called replication. A consequence of replication is that signals fed into the system at frequencies close to the sampling frequency, or a multiple of the sampling frequency will produce an output signal in the baseband (somewhere between dc and $f_s/2$). Such signals are indistinguishable from signals fed in close to dc. This typically unwanted output signal is called an aliased component.

When sampling is accompanied by a hold function, the frequency response is first convolved with the delta function to represent the sampling function. Then the result is multiplied by the hold function. The hold function can be shown to have a transfer function of:

$$G_{\text{hold}}(j\omega) = \frac{1 - e^{-j\omega T}}{j\omega} \frac{e^{j\omega T/2}}{e^{j\omega T/2}} = Te^{-j\omega T/2} \frac{e^{j\omega T/2} - e^{-j\omega T/2}}{2j\,\omega T/2}$$

$$= Te^{-j\omega T/2} \frac{\sin(\omega T/2)}{(\omega T/2)} \tag{5.63}$$

The effect of the hold function in general is to attenuate higher-frequency components in the output spectrum.

Thus, from the sampling theory, it is known that the frequency spectrum of a sampled system repeats once every sampling frequency, and therefore the spectrum of the quantization noise in a sampled system will be centered around dc and spread out to half of the sampling frequency $f_s/2$ and there will be a copy of the noise spectrum from $f_s/2$ to $3f_s/2$ and so on. Considering that all the noise power lies in the

Figure 5.43   Impulses in the time-domain become frequency-domain pulses.

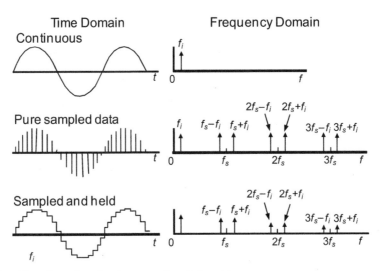

Figure 5.44 The effect of replication on an input signal f_i.

range of positive frequency band, that is, $0 \leq f \leq \infty$, the quantization noise power thus folds into the band from dc to $f_s/2$. Assuming white noise, the power spectral density $E^2(f)$ of the quantization noise is given by

$$E^2(f) = \frac{e^2_{rms}}{f_s/2} = 2Te^2_{rms} \tag{5.64}$$

where the sample period $T = 1/f_s$. For a band-limited signal $0 \leq f < f_0$ with bandwidth of f_0, the quantization noise power that falls into the signal band can be found as

$$n_0^2 = \int_0^{f_0} E^2(f)df = 2f_0Te^2_{rms} = \frac{\Delta^2 f_0}{6 \cdot f_s} = \frac{\Delta^2}{12 \cdot OSR} \tag{5.65}$$

where the *oversampling rate* (OSR) is defined as the ratio of the sampling frequency f_s to the Nyquist frequency $2f_0$, that is

$$OSR = \frac{f_s}{2f_0} \tag{5.66}$$

In an N-bit sampled system, if the quantizer has 2^N quantization levels equally spaced by Δ, then the maximum peak-to-peak amplitude is given by

$$v_{max} = \left(2^N - 1\right) \cdot \Delta \tag{5.67}$$

This equation is an approximation, but if the signal is sinusoidal and if N is large the approximation will be a small one. The associated signal power is

$$P = \frac{1}{8}\left(2^N - 1\right)^2 \cdot \Delta^2 \tag{5.68}$$

Thus, the dynamic range (DR) (the difference between highest input level and the noise floor) by making use of (5.65) becomes

$$DR = 10\log\left(\frac{\frac{1}{8}\left(2^N - 1\right)^2 \Delta^2}{n_0^2}\right) \approx 10\log\left(\frac{3 \cdot 2^{2N}\,OSR}{2}\right) \qquad (5.69)$$

Noting that $\log_{10}(x) = \log_{10}(2)\cdot\log_2(x)$, the above expression becomes:

$$DR \approx 6.02 \cdot N + 3 \cdot \log_2(OSR) + 1.76 \qquad (5.70)$$

Therefore, the DR improves by 6 dB for every bit added to the quantizer. For the same amount of total quantization noise power, every doubling of the sampling frequency reduces the in-band quantization noise by 3 dB. Note that while ADCs have been talked about here, this result would hold for a DAC as well.

The other main source of noise in an ADC or DAC is the timing jitter of the reference clock. The rms timing jitter t_{jitter} is related to the integrated phase noise of the of the reference clock by

$$t_{jitter} = \frac{IntPN_{rms}}{2\pi} \cdot T_{clk} \qquad (5.71)$$

where T_{clk} is the period of the reference clock. Now if the input to the ADC or the output of a DAC is assumed to be a sine wave at a frequency f_{in}, the amount of noise caused by reference clock jitter can be estimated. If the input waveform is given by

$$v_{in}(t) = A\sin\left(2\pi \cdot f_{in}t\right) \qquad (5.72)$$

then the slope of the input is simply

$$\frac{dv_{in}(t)}{dt} = 2\pi \cdot f_{in}A\cos\left(2\pi \cdot f_{in}t\right) \qquad (5.73)$$

This slope will be highest at the waveform zero crossings and zero at the waveform peak and if there is an error in the time the waveform is sampled t_{jitter}, then there will be an rms error in the sampled voltage of

$$v_{in\_error\_rms} = 2\pi \cdot \frac{f_{in}A}{\sqrt{2}} \cdot t_{jitter} \qquad (5.74)$$

Note that the root 2 comes from making this an rms value.

In this case the SNR will be given by [taking the ratio of (5.74) to the amplitude of (5.72), which is A]

$$SNR_{jitter} = -20\log\left(2\pi \cdot f_{in} \cdot t_{jitter}\right) \qquad (5.75)$$

Therefore, it is now possible to estimate the number of bits and the clock jitter required to assure that the DR of the ADC or DAC does not degrade the overall system SNR by more than the required amount.

Example 5.18:   Specifying an ADC

An ADC is required to have a SNR of 30 dB and the signal bandwidth is 20 MHz. Ignoring any out-of-band interferers give as much detail as possible about the required performance of the ADC.

Solution:

First let us assume that we are going to use a Nyquist rate ADC for this design and therefore the ADC will be clocked at 40 MHz and the OSR will be one. To give some margin in the design, we will calculate the number of bits required for a SNR of 33 dB. In this case, making use of (5.70), this ADC will need

$$N = \frac{SNR - 1.76}{6.02} = 5.2$$

Thus, in this case a 6-bit ADC will be adequate. Next we can calculate the requirement on the reference clock jitter. Again assuming 33 dB to give some margin

$$t_{jitter} = \frac{\log^{-1}\left(\dfrac{-SNR_{jitter}}{20}\right)}{2\pi \cdot f_{in}} = 0.18 \text{ ns}$$

Therefore, this application will require a reference clock with better than 0.18 ns of jitter.

References

[1]   Razavi, B., *RF Microelectronics*, Upper Saddle River, NJ: Prentice Hall, 1998.

[2]   Crols, J., and M. Steyaert, *CMOS Wireless Transceiver Design*, Dordrecht, the Netherlands: Kluwer Academic Publishers, 1997.

[3]   Couch II, L. W., *Digital and Analog Communication Systems*, 6th ed., Upper Saddle River, NJ: Prentice Hall, 2001.

[4]   Carson, R. S., *Radio Communications Concepts: Analog*, New York: John Wiley & Sons, 1990.

[5]   Rohde, U. L., J. Whitaker, and A. Bateman, *Communications Receivers: DSP, Software Radios, and Design*, 3rd ed., New York: McGraw-Hill, 2000.

[6]   Larson, L. E., (ed.), *RF and Microwave Circuit Design for Wireless Communications*, Norwood, MA: Artech House, 1997.

[7]   Gu, Q., *RF System Design of Transceivers for Wireless Communications*, 1st ed., New York: Springer Science and Business Media, 2005.

[8]   Sheng, W., A. Emira, and E. Sanchez-Sinencio, "CMOS RF Receiver System Design: A Systematic Approach," *IEEE Transactions on Circuits and Systems I: Regular Papers*, Vol. 53, No. 5, May 2006, pp. 1023–1034.

[9]   Rappaport, T. S., *Wireless Communications: Principles and Practice*, 2nd ed., Upper Saddle River, NJ: Prentice-Hall, 2001.

[10] McCune, E., *Practical Digital Wireless Signals*, Cambridge, U.K.: Cambridge University Press, 2010.

CHAPTER 6

Frequency Synthesis

6.1 Introduction

There are many ways to realize synthesizers, but possibly the most common is based on a phase-locked loop (PLL). PLL-based synthesizers can be further subdivided by which type of a division is used in the feedback path. The division ratio N can be either an integer or a fractional number. If the number is fractional, then the synthesizer is called a *fractional-N synthesizer*. This type of synthesizer can be further distinguished by the method used to control the divide ratio, for example, by a sigma-delta controller or by some other technique. In this chapter, analysis is done with a general N without details of how N is implemented; thus, the analysis is applicable both to integer-N and fractional-N synthesizers. There will also be discussion on all-digital synthesizers for which there may no longer be a traditional divider in the loop. Additional general information on synthesizers can be found in [1–53].

6.2 Integer-*N* PLL Synthesizers

An integer-N *PLL* (phase-locked loop) is the simplest type of phase-locked loop synthesizer and is shown in Figure 6.1. Note that N refers to the divide-by-N block in the feedback of the PLL. The two choices are to divide by an integer, (integer N), or to divide by a fraction (fractional N), essentially by switching between two or more integer values such that the effective divider ratio is a fraction. This is usually accomplished by using a $\Sigma\Delta$ modulator to control the division ratio. PLL-based synthesizers are amongst the most common ways to implement a synthesizer and this area is the subject of a great deal of research and development [20–52]. The PLL based synthesizer is a feedback system that compares the phase of a reference f_r to the phase of a divided down output of a controllable signal source f_{fb} (also known as a *voltage-controlled oscillator* or a VCO). The summing block in the feedback is commonly called a phase detector. Through feedback, the loop forces the phase of the signal source to track the phase of the feedback signal and therefore their frequencies must be equal. Thus, the output frequency, which is a multiple of the feedback signal, is given by

$$f_o = N \cdot f_{\text{ref}} \tag{6.1}$$

Due to divider implementation details, it is not easily possible to make a divider that divides by non-integer values. Thus, a PLL synthesizer of this type is called an integer-N frequency synthesizer. Circuits inside the feedback loop can be described by their transfer functions. These transfer functions can be designed to engineer the system dynamics to meet design specifications for the synthesizer. Typically, $F(s)$ is realized with a *lowpass filter* (LPF) or lowpass network. The details of the loop and the circuit components in the loop will be discussed later in this chapter.

In brief, the PLL is a feedback system which forces the divided-down VCO output phase to follow the reference signal phase. That is, it is a negative feedback loop with phases as the input and output signals. The loop is composed of a phase detector, a lowpass filter, a VCO, and a divider. The phase detector, which is the summing block of the feedback system, is used to compare output phase θ_o to reference phase θ_R. The lowpass filter is usually a linear transfer function that is placed in the system to control the loop's dynamic behavior including the settling time and transient response. The VCO generates the output signal and the divider divides the VCO output signal back down to the same frequency as the input. We use a feedback system based on phase rather than frequency because in any feedback loop without infinite dc gain there is always an error (finite error signal) between the input (reference) and the output. Thus, if we used a frequency-locked loop, then, in most cases, there would be an error in the output frequency and it would not track the input as precisely as does a loop based on phase. The input reference in wireless communications is a quartz crystal. These crystals are low cost and can be made to resonate with extreme accuracy at a particular frequency determined by the physical properties of the crystal. Unfortunately, they can only resonate at frequencies as high as about 100 MHz and therefore cannot be used directly as an LO in RF applications. The other disadvantage to using a crystal directly is that there is no convenient way to tune its frequency. This is one of the main reasons that frequency synthesizers have become so popular. If the divider block is implemented using circuitry such that the divide ratio is programmable, then a range of frequencies can be obtained without the need to change the reference frequency.

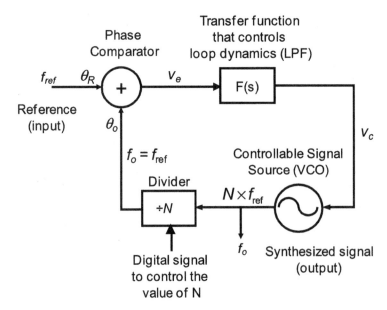

Figure 6.1 A simple integer-*N* frequency synthesizer.

Because N is an integer, the minimum step size of this synthesizer is equal to the reference frequency f_r. Therefore, to get a smaller step size, the reference frequency must be made smaller. This is often undesirable so instead a fractional-N design is often used. This will be discussed later.

6.3   PLL Components

We will now briefly look at the basic components needed to make a PLL-based synthesizer and their basic governing equations. Only a very basic introduction to these components will be given in this chapter and only the most common forms of these circuits will be considered here.

6.3.1   Voltage-Controlled Oscillators (VCOs) and Dividers

At the most basic level, all VCOs will have an output frequency with some dependence on the control voltage (or sometimes control current) as shown in Figure 6.2. Note that the curve is not always linear (actually it is hardly ever linear, but for the moment we will assume that it is). Also note that the control voltage can usually be adjusted between ground and the power supply voltage and that, over that range, the VCO frequency will move between some minimum and some maximum value.

Here V_c is the nominal control voltage coming from the loop filter, and ω_{nom} is the nominal frequency at this nominal voltage. Usually when considering loop dynamics, we only consider frequency deviations away from the nominal frequency ω_{VCO} and voltage deviations away from the nominal voltage v_c. Thus, we can write the oscillating frequency as

$$\omega_o = \omega_{\text{nom}} + \omega_{\text{VCO}} = \omega_{\text{nom}} + K_{\text{VCO}} v_c \qquad (6.2)$$

where

$$v_c = v_C - V_{C\_\text{nom}} \qquad (6.3)$$

In addition, if we remove this "dc" part of the equation and only consider the part that is changing, we are left with

$$\omega_{\text{VCO}} = K_{\text{VCO}} v_c \qquad (6.4)$$

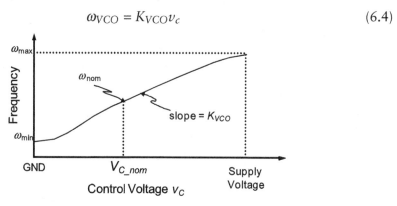

Figure 6.2   A typical VCO characteristic.

However, we would like to have an expression relating input voltage to output phase since the output of the VCO ultimately goes to the phase detector. To relate frequency ω to phase θ, we note that

$$\omega = \frac{d\theta}{dt} \tag{6.5}$$

Therefore, the output phase of the VCO can be given as

$$\theta_{VCO} = \int \omega_{VCO} dt = K_{VCO} \int_0^t v_c(t) dt \tag{6.6}$$

In the Laplace domain, this becomes

$$\frac{\theta_{VCO}(s)}{v_c(s)} = \frac{K_{VCO}}{s} \tag{6.7}$$

Thus, we have the desired equation for the transfer function of the VCO block. Note that, for the purposes of system behavior, the divider can be thought of as an extension of the VCO. The output phase after the divider is simply

$$\frac{\theta_o}{v_c} = \frac{1}{N} \cdot \frac{K_{VCO}}{s} \tag{6.8}$$

6.3.2 Phase Detectors

A phase detector produces an output signal proportional to the phase difference of the signals applied to its inputs. The inputs and outputs can be sine waves, square waves, or other periodic signals, not necessarily having a 50% duty cycle. The output signal could be a current or voltage and it could have multiple frequency components. As the dc value is the component of interest, the phase detector is typically followed by some sort of filter. Thus, the equation that describes a phase detector is

$$v_e(s) = K_{\text{phase}} \left(\theta_R(s) - \theta_o(s) \right) \tag{6.9}$$

provided that the phase detector is a continuous time circuit (which is often not the case in integrated circuit implementations). The output of the phase detector, $v_e(s)$, is often also called the error voltage and is seen to be proportional to the difference of the input phases with proportionality constant K_{phase}. This is a linearized equation, often valid only over limited range. Another comment that can be made about phase detectors is that often they respond in some way to a frequency difference as well. In such a case, the circuit is often referred to as a *phase-frequency detector* (PFD).

The phase detector can be as simple as an XOR gate or an XNOR gate, but typical phase detectors are usually more complicated. For example, flip-flops form the basis of tri-state phase detectors, which is often combined with a charge pump. This type of phase detector has two outputs as shown in Figure 6.3. If the reference phase (v_R) is ahead of the output phase (v_o), then the circuit produces a signal UP that tells the VCO to speed up and therefore advance its phase to catch up with the reference phase. Conversely, if the reference phase is lagging the output phase, it produces a signal DN that tells the VCO to slow down and therefore retard its phase to match the reference phase. If the reference and output are in phase, then the phase detector does not produce an output. The signals UP and DN are sometimes also called v_U and v_D, respectively.

The two digital signals produced by a PFD have to be converted back into an analog control signal at the input of the VCO, and the circuit most commonly used to do this is called a charge pump. A charge pump is made of two controllable current sources connected to a common output also shown in Figure 6.3. The outputs from the phase detector turn on one of the two currents, which either charge or discharge capacitors attached to the VCO input.

A simple implantation of the PFD is shown in Figure 6.4(a), and a description of the PFDs operation based on the state diagram shown in Figure 6.4(b). Transitions happen only on the rising edge of v_o or v_R. Let us assume we start in the middle state, the tri-state where both outputs are zero. Then, depending on which edge arrives first, the PFD moves either to the Up state or to the Down state. If the reference edge arrives first, the output needs to catch up and so the up switch turns on to charge up the output. It stays up until the other edge comes along; thus, the average output current from the charge pump depends on how far apart the two signals are. However, if the reference is behind the output, then the output is too fast and needs to be slowed down. This causes a down pulse to be generated, and as a result current flows out of the charge pump, discharging the capacitors. Current flows for the length of time τ between the output edge and the reference edge. If the period is T, the average output current is

$$i_d = I\frac{\tau}{T} = \left(\frac{I}{2\pi}\right)(\theta_R - \theta_o)$$ (6.10)

Thus K_{phase} for this phase detector is

$$K_{\text{phase}} = \frac{I}{2\pi}$$ (6.11)

where I is the current that flows through the controllable current sources in the charge pump when they are on.

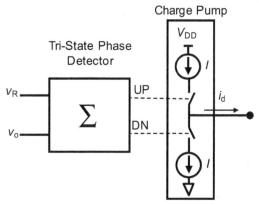

Figure 6.3 Tri-state phase detector and charge pump.

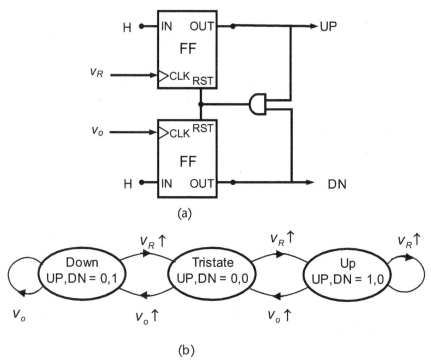

(a)

(b)

Figure 6.4   (a) Simple PFD implementation, and (b) PFD state diagram.

The operation of the PFD and current sources is shown in Figure 6.5. The movement from state to state is controlled by the rising edge only, so the pulse width of the input signals is not important. We have shown the pulses as being narrow, but it would work equally well using wider pulses. In the figure, it is shown that for the same phase difference between v_o and v_R, the output current depends on where the operation starts. In Figure 6.5(a), the v_o edge comes first, resulting in down pulses and a negative average output current. In Figure 6.5(b), the v_R edge comes first, resulting in up pulses and a positive average output current. We note also that if the v_o pulse was delayed (moved towards the right), in Figure 6.5(a), the down pulses would become narrower resulting in average current closer to zero. In Figure 6.5(b), for the same delay of the v_o pulses, the up pulses become wider and the average current moves closer to I. Note that the short pulses associated with UP in Figure 6.5(a) and DN in Figure 6.5(b) are realistic. They are the result of details of the PFD design.

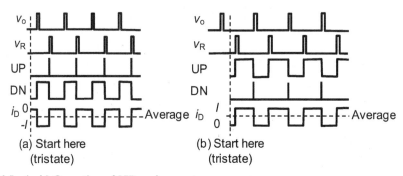

Figure 6.5   (a, b) Operation of PFD and current sources.

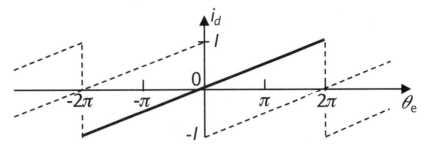

Figure 6.6 Average output current versus phase for PFD and charge pump.

If this average output current is now plotted as a function of the phase difference, with v_R taken as the reference, the result can be interpreted as the transfer function of the phase detector and charge pump and is shown in Figure 6.6. We note that any positive phase (for example, 60°, for which the output current would be $I \times$ 60/360) could be interpreted as the equivalent negative phase (for example, +60° is equivalent to –300° for which the current is equal to $-I \times 300/360$). Thus, for every phase difference, there are two possible interpretations and this is shown by the solid and dashed lines in Figure 6.6. We note that this is equivalent to starting the phase detector in a different state or at a different time, as was shown in Figure 6.5.

To illustrate how this phase detector can also be used as a frequency detector, in Figure 6.7, waveforms are shown for two different input frequencies. We have assumed that we start in tri-state. As the output pulse v_o occurs first, down pulses DN occur, which would result in negative output current pulses i_d and an average negative output current, shown by the dotted line. However, as the frequencies are different, the output pulse width is changing, in this case becoming narrower and the average current is moving towards zero. Eventually, the phase detector experiences a second reference pulse v_R before the output pulse v_o and moves into the up state, and up current pulses i_d result. From then on, the phase detector output states will be either tri-state or in the up state, so only positive current is ever provided. In this way, it can be seen that for a reference frequency higher than the output frequency, average current is positive. Similarly, for a reference frequency lower than the output frequency, the average output current would always be negative (except, of course, for a possible short time at startup). Thus, with the correct feedback polarity, this current can be used to correct the VCO frequency until it is the same as the reference frequency and the loop is locked.

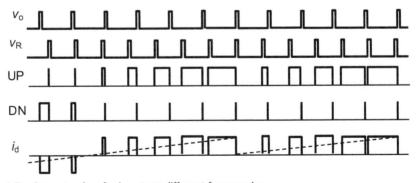

Figure 6.7 Output pulses for inputs at different frequencies.

6.3.3 The Loop Filter

Normally, VCOs are controlled by voltage and not current. Thus, generally, we need a method to turn the current produced by the charge pump back into a voltage. In addition, lowpass filtering is needed since it is not desirable to feed pulses into the VCO. This is usually done by dumping the charge produced by the charge pump onto the terminals of a capacitor. As we will show later, a simple capacitor all by itself does not yield a stable loop, so a combination of capacitors and resistors are used. This part of the PLL is typically called the loop filter. One of the most common loop filters used is shown in Figure 6.8. Note that PLLs that do not use charge pumps have loop filters as well. In addition to turning the current back into the voltage, loop filters are also the components most commonly used to control system-level loop dynamics.

The frequency response of the phase frequency detector, charge pump, and loop filter is mainly determined by the loop filter. The frequency response of the network (seen in Figure 6.8) will now be analyzed, starting with the admittance of the capacitor and resistor circuit

$$Y = sC_2 + \frac{1}{R + \frac{1}{sC_1}} = sC_2 + \frac{sC_1}{sC_1 R + 1} = \frac{sC_2(sC_1 R + 1) + sC_1}{sC_1 R + 1} \tag{6.12}$$

This admittance can be used to determine v_c the control voltage of the VCO using (6.11):

$$v_c = \frac{i_d}{Y} = \frac{K_{\text{phase}}(\theta_R - \theta_o)(sC_1 R + 1)}{sC_2(sC_1 R + 1) + sC_1} = \frac{K_{\text{phase}}(\theta_R - \theta_o)(1 + sC_1 R)}{s(C_1 + C_2)(1 + sC_s R)} \tag{6.13}$$

where $C_s = \dfrac{C_1 C_2}{C_1 + C_2}$ and $K_{\text{phase}} = \dfrac{I}{2\pi}$.

The frequency response of the charge pump and filter as given in (6.13) is shown in Figure 6.9. We note that at low frequencies, the response is dominated by the pole at the origin in the transfer function and thus the circuit acts like an integrator. Also note that this has been derived in continuous time and, as long as the pulses are much faster than any changes of interest at the output, this is a reasonable assumption.

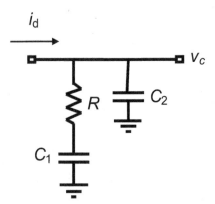

Figure 6.8 A typical loop filter.

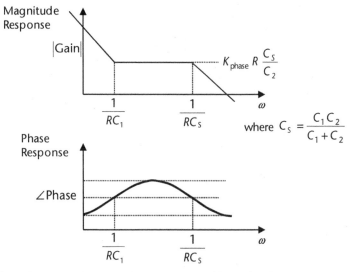

Figure 6.9   Phase-frequency detector, charge pump, and loop filter frequency response.

6.4   Continuous-Time Analysis for PLL Synthesizers

The s domain model for a synthesizer is shown in its most general form in Figure 6.10. Here any loop filter (in combination with a charge pump) is simply shown as $F(s)$, and the dc gain is brought out explicitly as term A_o. We can therefore derive the basic loop transfer function for the loop.

6.4.1   Simplified Loop Equations

The overall transfer function is

$$\frac{\theta_o}{\theta_R} = \frac{\dfrac{A_o K_{\text{phase}} F(s)}{N} \cdot \dfrac{K_{VCO}}{s}}{1 + \dfrac{A_o K_{\text{phase}} F(s)}{N} \cdot \dfrac{K_{VCO}}{s}} = \frac{KF(s)}{s + KF(s)} \tag{6.14}$$

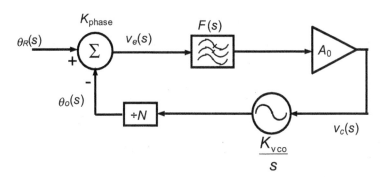

Figure 6.10   Complete loop in the frequency domain.

where K is given by

$$K = \frac{A_o K_{\text{phase}} K_{VCO}}{N} \tag{6.15}$$

Now (6.14) is the most general PLL loop equation, and for specific loops it differs only in the form that $F(s)$ takes. For instance, in a first-order loop $F(s)$ is simply equal to 1. In this case, the loop equation becomes

$$\frac{\theta_o}{\theta_R} = \frac{K}{s + K} \tag{6.16}$$

Note that for this first-order loop, for a phase ramp (change in frequency), the phase error is not zero because there are not enough integrators in the loop. As zero phase error is often highly desired and due to a lack of flexibility, this loop is not often used in practice.

A much more common PLL is the second-order PLL, an example of which is the charge-pump and PFD-based PLL. A typical second-order PLL has a loop filter with a transfer function of

$$F(s) = \frac{\tau s + 1}{s} \tag{6.17}$$

A charge-pump and PFD-based PLL with the loop filter, as previously discussed, is an example of a second-order PLL as will now be shown. The most common system level configuration is shown in Figure 6.11. In this case (assuming for the moment that we ignore C_2), the impedance of the loop filter is

$$F(s) = R + \frac{1}{sC_1} = \frac{sC_1 R + 1}{sC_1} \tag{6.18}$$

Note this transfer function is an impedance, because this stage converts current to voltage, which is not a typical transfer function, but works here. Thus, we can substitute this back into (6.14) and therefore find

$$\frac{\theta_o}{\theta_R} = \frac{\dfrac{IK_{VCO}}{2\pi \cdot N}\left(R + \dfrac{1}{sC_1}\right)}{s + \dfrac{IK_{VCO}}{2\pi \cdot N}\left(R + \dfrac{1}{sC_1}\right)} = \frac{\dfrac{IK_{VCO}}{2\pi \cdot NC_1}(RC_1 s + 1)}{s^2 + \dfrac{IK_{VCO}}{2\pi \cdot N}Rs + \dfrac{IK_{VCO}}{2\pi \cdot NC_1}} \tag{6.19}$$

Thus, for this PLL, we get a second-order transfer function with a zero. Note that the purpose of R can be seen directly from this equation. If R is set equal to zero, it can be seen by inspection of (6.19) that the poles of this equation will sit on the $j\omega$ axis and the loop would oscillate or be on the verge of oscillating. From (6.19), expressions for the loop dynamics can be determined. The natural frequency of the loop is given by

$$\omega_n = \sqrt{\frac{IK_{VCO}}{2\pi \cdot NC_1}} \tag{6.20}$$

The damping constant is given by

$$\zeta = \frac{R}{2}\sqrt{\frac{IK_{VCO}C_1}{2\pi \cdot N}} \tag{6.21}$$

Often the resistor and capacitor value are to be determined for a known damping constant and natural frequency. It is straightforward to solve these two equations for these variables:

$$C_1 = \frac{IK_{VCO}}{2\pi \cdot N\omega_n^2} \tag{6.22}$$

and

$$R = 2\zeta\sqrt{\frac{2\pi \cdot N}{IK_{VCO}C_1}} = \zeta\frac{4\pi \cdot N\omega_n}{IK_{VCO}} \tag{6.23}$$

From the above, it can be shown that (6.19) can be rewritten in a general form as

$$\frac{\theta_o}{\theta_R} = \frac{\omega_n^2\left(\dfrac{2\zeta}{\omega_n}s+1\right)}{s^2+2\zeta\omega_n s+\omega_n^2} \tag{6.24}$$

This shows that there is a relationship between the pole and zero locations.

Note that it is easy to determine the transfer function even if the output is taken from other places in the loop. For instance, it is often interesting to look at the control voltage going into the VCO. In this case, the system transfer function becomes

$$\frac{v_c}{\theta_R} = \frac{\dfrac{I\cdot s}{2\pi \cdot C_1}(RC_1s+1)}{s^2+\dfrac{IK_{VCO}}{2\pi \cdot N}Rs+\dfrac{IK_{VCO}}{2\pi \cdot NC_1}} = \frac{\dfrac{N\omega_n^2}{K_{VCO}}s\left(\dfrac{2\zeta}{\omega_n}s+1\right)}{s^2+2\zeta\omega_n s+\omega_n^2} \tag{6.25}$$

This expression contains an extra s in the numerator. This makes sense because the control voltage is proportional to the frequency of the VCO, which is the derivative of the phase of the output. Note that we can also write an expression for the output frequency (as given by the control voltage) as a function of the input frequency, noting that frequency is the derivative of phase and starting from (6.25)

$$\frac{V_C}{\theta_R s} = \left[\frac{\dfrac{I\cdot s}{2\pi \cdot C_1}(RC_1s+1)}{s^2+\dfrac{IK_{VCO}}{2\pi \cdot N}Rs+\dfrac{IK_{VCO}}{2\pi \cdot NC_1}}\right]\frac{1}{s} = \left[\frac{\dfrac{N\omega_n^2}{K_{VCO}}s\left(\dfrac{2\zeta}{\omega_n}s+1\right)}{s^2+2\zeta\omega_n s+\omega_n^2}\right]\frac{1}{s}$$

$$\frac{V_C}{\omega_R} = \frac{\dfrac{N\omega_n^2}{K_{VCO}}\left(\dfrac{2\zeta}{\omega_n}s+1\right)}{s^2+2\zeta\omega_n s+\omega_n^2} \tag{6.26}$$

which is nearly identical to the expression for phase transfer function θ_o/θ_R given by (6.24).

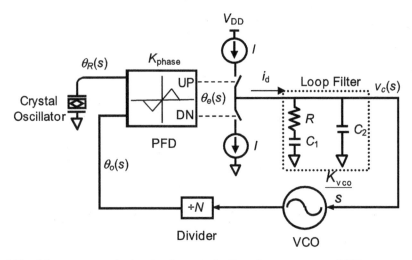

Figure 6.11 A frequency synthesizer implemented with a charge pump and PFD.

6.4.2   PLL System Frequency Response and Bandwidth

Figure 6.12 is a plot of the closed loop transfer function for the charge pump and PFD based PLL, which is described by (6.19) and (6.24) for different values of damping constant. This diagram shows that the loop's 3-dB bandwidth is highly dependent on the damping constant. It can be shown that the 3-dB bandwidth of this system is given by [2]

$$\omega_{3\text{dB}} = \omega_n\sqrt{1+2\zeta^2+\sqrt{4\zeta^4+4\zeta^2+2}} \tag{6.27}$$

Because this equation can be tedious without a calculator, two equations sometimes used to approximate this are

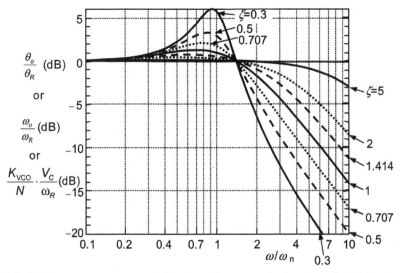

Figure 6.12 PLL frequency response of the closed loop transfer function of a high-gain second-order loop. Note that the graph is valid for any of the three functions shown on the y-axis.

$$\omega_{3dB} \approx 2\zeta\omega_n \quad \zeta > 1.5 \text{ (approximation \#1)}$$

$$\omega_{3dB} \approx \left(1 + \zeta\sqrt{2}\right)\omega_n \quad \zeta < 1.5 \text{ (approximation \#2)} \qquad (6.28)$$

6.4.3   Complete Loop Transfer Function Including C_2

Note that if C_2 is included, this adds a high frequency pole to the system. Normally, this capacitor is chosen to be about one-tenth of the value of C_1 and is included to clean up high-frequency ripple on the control line. If this capacitor is included, then the following expression for open-loop gain can be derived

$$\left(\frac{\theta_o}{\theta_R}\right)_{\text{open loop}} = \frac{K_{\text{VCO}}K_{\text{phase}}\left(1 + sC_1R\right)}{s^2 N\left(C_1 + C_2\right)\left(1 + sC_sR\right)} \qquad (6.29)$$

Here C_s is the series combination of C_1 and C_2. This is plotted in Figure 6.13, which shows a low-frequency slope of -40 dB/decade and $180°$ of phase shift. After the zero, the slope is -20 dB per decade and the phase heads back towards $90°$ of phase shift. After the high-frequency pole, the slope again is -40 dB/decade and the phase approaches $180°$. Note that the dashed lines in the figure show the response of the system if the capacitor C_2 is not included. For optimal stability (maximum phase margin in the system), the unity gain point should be at the geometric mean of the zero and the high-frequency pole, as this is the location where the phase shift is furthest from $180°$. Some may wonder, after seeing this plot, if the system is actually unstable at dc because, at this point, the phase shift is $180°$ and the gain is greater than 1. In fact, this is not the case and the system is stable. A full stability analysis, for example by plotting the closed-loop poles would show this.

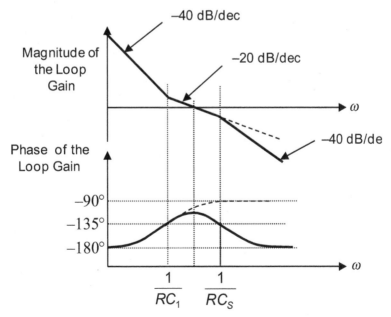

Figure 6.13   Open-loop magnitude and phase response. Note that the dotted line shows response if high-frequency pole is not included.

The closed-loop gain with C_2 is given by

$$\frac{\theta_o}{\theta_R} = \frac{K_{\text{VCO}}K_{\text{phase}}\left(1+sC_1R\right)}{s^2N\left(C_1+C_2\right)\left(1+sC_sR\right)+K_{\text{VCO}}K_{\text{phase}}\left(1+sC_1R\right)} \tag{6.30}$$

Thus, one could now estimate all the parameters of the loop. Figure 6.13 shows that if the zero and the high-frequency pole are relatively far apart, then up to the unity gain point, the loop parameters are nearly the same whether or not the high-frequency pole is included. There is, however, a slight decrease of phase margin (in the diagram from about 75° to about 65°).

A cautionary note about the choice of C_2 should also be mentioned now. It may not always be appropriate to use a value of one-tenth C_1 for the value of C_2. It has been assumed in this section that most of the time C_2 does not significantly change the loop dynamics; however, R increases at high ζ, and in this case the impedance of the series combination of C_1 and R may become comparable to the impedance of C_2 at the loop natural frequency. If this happens, then the equations derived in this section will become increasingly inaccurate, in which case there would be no choice but to reduce the value of C_2 to less than $C_1/10$.

Note that while in practice C_2 is almost always present, the results in the previous sections can be applied almost always to a third-order loop with a very small error. Thus, these results are also valid for this more practical case.

6.5 Discrete Time Analysis for PLL Synthesizers

The preceding linear continuous time analysis of the synthesizer is not valid under all conditions. If the loop bandwidth is increased so that it becomes a significant fraction of the reference frequency, the previous analyses become increasingly inaccurate. For this reason, it is sometimes necessary to treat the synthesizer system as a discrete time control system as it truly is [19]. To do this, we must consider the PFD, which in this case is the sampling element. We assume that in lock the PFD produces only narrow impulses at the reference frequency. Note that because the charge pump behaves like an integrator, it has infinite gain at dc. Thus, as long as the frequency deviation is small, this high gain drives the phase error towards zero and the output pulses will be narrow. Therefore, it acts like an ideal sampler. The loop filter holds the charge dumped onto it in each cycle so, in this system, it acts as a hold function. Hence, this is a sampled system, and the s domain combination of the VCO, divider, PFD, and the loop filter must be converted into their z domain equivalents (being careful to remember to multiply the functions together before converting them into the z domain). Thus, the closed-loop transfer function including the sampling action of these four blocks is (ignoring C_2) (see Figure 6.14) [2]

$$G(z) = \frac{K\left(z-\alpha\right)}{z^2+\left(K-2\right)z+\left(1-\alpha K\right)} \tag{6.31}$$

where

$$\alpha = \frac{4\zeta - \omega_n T}{4\zeta + \omega_n T} \tag{6.32}$$

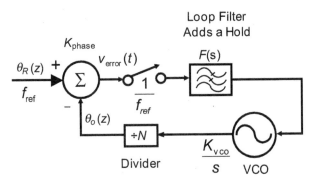

Figure 6.14   Discrete-time, system-level diagram for a synthesizer.

$$K = \frac{\omega_n^2 T^2}{2}\left(1 + \frac{4\zeta}{\omega_n T}\right) \tag{6.33}$$

and T is the period of the reference.

Starting from the closed loop transfer function, the pole locations as a function of the reference period T can be sketched and are shown in Figure 6.15. Note that depending on the specific parameters, this plot will change slightly, but the basic shape of the root locus will remain the same. The point of greatest interest here is that point at which the reference period is increased to some critical value and the system becomes unstable, as one of the poles moves outside the unit circle. At or even close to this period the s domain analysis discussed in the previous section will be increasingly inaccurate. Note that the reference frequency would not normally be this low in the design of a PLL, but the designer must be aware of this so that the assumptions of the previous section are not violated.

Now the poles of (6.31) are given by

$$\text{Poles} = 1 - \frac{K}{2} \pm \frac{1}{2}\sqrt{(K-2)^2 - 4(1 - \alpha K)} \tag{6.34}$$

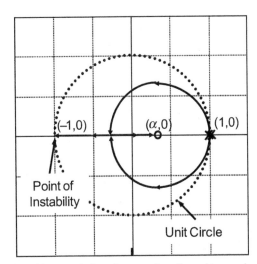

Figure 6.15   Sketch of closed-loop pole location as a function of increasing reference period.

The pole that has the larger positive value is not of concern, because it will never leave the unit circle. However, it is of interest to find out when

$$1 - \frac{K}{2} - \frac{1}{2}\sqrt{(K-2)^2 - 4(1-\alpha K)} = -1 \qquad (6.35)$$

Skipping a number of steps, (primarily because they are boring) this will happen when

$$K(1+\alpha) = 4 \qquad (6.36)$$

Taking this expression and substituting back in for K and α, the critical period for which the loop will go unstable, T_{US} is

$$T_{US} = \frac{1}{\omega_n \zeta} \qquad (6.37)$$

Noting that

$$T_{US} = \frac{2\pi}{\omega_{ref\_crt}} \qquad (6.38)$$

where ω_{ref_crt} is the reference frequency at which the loop goes unstable, it can be determined that

$$\omega_{ref\_crt} = 2\pi\zeta\omega_n \qquad (6.39)$$

Therefore,

$$\frac{\omega_{ref}}{\omega_n} \geq 2\pi\zeta \qquad (6.40)$$

So, for instance, in the case of $\zeta = 0.707$, this ratio must be greater than 4.4. Therefore, for a reference frequency of 40 MHz, if the loop natural frequency is set any higher than 9.1 MHz, the loop will go unstable. A number that is quoted often as being a "safe" ratio is 10 to 1 [15].

6.6   Transient Behavior of PLLs

In the two previous sections, linear s domain equations that describe the PLL as a classic feedback system, and more complicated z domain equations were derived. However, the behavior of a real PLL is much more complex than either of these two analyses can explain. This is because until the loop starts tracking the phase of the input, or alternatively if there is a very large step or ramp in the phase of the input, the loop's output phase may not be able to follow the input phase. This is primarily due to the limitations of the phase detector, which has a narrow linear range. For example, the tri-state PDF has a linear range of $\pm 2\pi$. If an event at the input occurs that causes the phase error to exceed 2π, then the loop will experience a nonlinear event: cycle slipping. Remember that in the previous analysis it was assumed that the phase detector was linear. This nonlinear event will cause a transient response that cannot be predicted by the theory of the previous section. The loop will, of course, work to correct the phase and force the VCO to track the input once more. When the loop goes into this process, it is said to be in acquisition mode as it is trying to acquire phase lock, but has not yet. Note that acquisition also happens when the PLL is first powered, because the VCO and reference will be at a random phase and probably will have a frequency difference. In extreme cases of this, the VCO may even be forced beyond its linear range of operation, which may result in the loop losing lock indefinitely. These situations will now be explored. First, the case where the loop is in lock, and experiences no cycle slipping will be considered.

6.6.1   PLL Linear Transient Behavior

Here the linear transient response of the most common PLL presented in this chapter will be considered further. In this section, only the s domain response will be discussed, which under most normal operating conditions is sufficient. However, the z domain equivalent of this analysis could be undertaken. For the linear transient behavior, the phase error rather than the output phase is needed, so a different transfer function has to be derived for the system shown in Figure 6.10. The result is

$$\frac{\theta_e}{\theta_R} = \frac{s^2}{s^2 + 2\zeta\omega_n s + \omega_n^2} \tag{6.41}$$

In this section, we will see the response to an input frequency step $\Delta\omega$. Because the input is described by phase, we take the phase equivalent of a frequency step, which is equivalent to a ramp of phase (we note that phase is the integral of frequency and the integral of a step is a ramp). Thus, the input is described by

$$\theta_R = \frac{\Delta\omega}{s^2} \tag{6.42}$$

This input in (6.42) when multiplied by the transfer function (6.41) results in

$$\theta_e = \frac{\Delta\omega}{s^2 + 2\zeta\omega_n s + \omega_n^2} \tag{6.43}$$

Then the inverse Laplace transform is taken, with the following results

$$\theta_e(t) = \frac{\Delta\omega}{\omega_n}\left[\frac{\sinh\omega_n\sqrt{\zeta^2 - 1}t}{\sqrt{\zeta^2 - 1}}\right]e^{-\zeta\omega_n t} \quad \zeta > 1 \tag{6.44}$$

$$\theta_e(t) = \frac{\Delta\omega}{\omega_n}\omega_n t \cdot e^{-\omega_n t} \quad \zeta = 1 \tag{6.45}$$

$$\theta_e(t) = \frac{\Delta\omega}{\omega_n}\left[\frac{\sin\omega_n\sqrt{1 - \zeta^2}t}{\sqrt{1 - \zeta^2}}\right]e^{-\zeta\omega_n t} \quad \zeta < 1 \tag{6.46}$$

These results are plotted in Figure 6.16 for various values of damping constant.

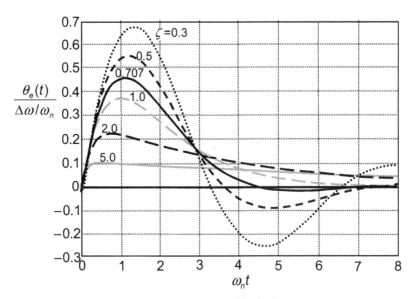

Figure 6.16 Error for frequency step, high-gain second-order loop. Note $\theta_e = \theta_R - \theta_0$.

It can be seen that a damping constant of 0.707 to 1 results in the fastest settling (reduction of phase error to zero). Depending on the required level of settling, one can determine the settling time. To the accuracy of the above diagram, settling is better than 99% complete when $\omega_n t = 7$ for $\zeta = 0.707$. Thus, given a required settling time, one can calculate the required natural frequency. To prevent the reference frequency from feeding through to the VCO, the loop bandwidth, as shown in Figure 6.12 and estimated in (6.28), must be significantly less than the reference frequency. In fact, the extra capacitor in the loop filter has been added to provide attenuation at the reference frequency.

It is also interesting to look at the control voltage:

$$\frac{V_C}{\omega_R} = \frac{\dfrac{N\omega_n^2}{K_{VCO}}\left(\dfrac{2}{\omega_n}\zeta s + 1\right)}{s^2 + 2\zeta\omega_n s + \omega_n^2} \tag{6.47}$$

In this case again we apply a step in frequency so that

$$\omega_R = \frac{\Delta\omega}{s} \tag{6.48}$$

Note that this equation is given in the frequency rather than phase domain, so s is raised to unity power in the denominator. Therefore, the control voltage is given by

$$V_C = \frac{\dfrac{N\omega_n^2}{K_{VCO}}\left(\dfrac{2\zeta}{\omega_n}s + 1\right)}{s^2 + 2\zeta\omega_n s + \omega_n^2} \cdot \frac{\Delta\omega}{s} = \frac{\dfrac{2\zeta \cdot N\omega_n}{K_{VCO}} \cdot \Delta\omega}{s^2 + 2\zeta\omega_n s + \omega_n^2} + \frac{\dfrac{N\omega_n^2}{K_{VCO}} \cdot \dfrac{\Delta\omega}{s}}{s^2 + 2\zeta\omega_n s + \omega_n^2} \tag{6.49}$$

Now the first term is simply a scaled version of the previous expression, and the second term is the integral of the first term. Therefore, the transient expression for the control voltage is given by

$$V_C(t) = \frac{\zeta \cdot N\Delta\omega}{K_{VCO}} \left[\frac{\sinh \omega_n \sqrt{\zeta^2 - 1}\, t}{\sqrt{\zeta^2 - 1}} \right] e^{-\zeta\omega_n t} - \frac{N\Delta\omega}{K_{VCO}} \left[\cosh \omega_n \sqrt{\zeta^2 - 1}\, t \right] e^{-\zeta\omega_n t}$$

$$+ \frac{N\Delta\omega}{K_{VCO}} \quad \zeta > 1 \tag{6.50}$$

$$V_C(t) = \frac{N\Delta\omega}{K_{VCO}} \omega_n t \cdot e^{-\omega_n t} - \frac{N\Delta\omega}{K_{VCO}} \cdot e^{-\omega_n t} + \frac{N\Delta\omega}{K_{VCO}} \quad \zeta = 1 \tag{6.51}$$

$$V_C(t) = \frac{\zeta \cdot N\Delta\omega}{K_{VCO}} \left[\frac{\sin \omega_n \sqrt{1 - \zeta^2}\, t}{\sqrt{1 - \zeta^2}} \right] e^{-\zeta\omega_n t} - \frac{N\Delta\omega}{K_{VCO}} \left[\cos \omega_n \sqrt{1 - \zeta^2}\, t \right] e^{-\zeta\omega_n t}$$

$$+ \frac{N\Delta\omega}{K_{VCO}} \quad \zeta < 1 \tag{6.52}$$

These expressions are plotted in Figure 6.17. It is interesting to note that, from this expression, it looks like high ζ is best for fast settling; however, it should be noted that even though the frequency appears to lock quickly, there is still a long period before the system is phase locked. Therefore, although these plots may be useful to compare with the control voltage (which is more readily available from simulation), they can also be misleading.

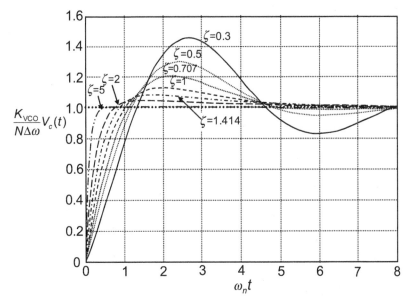

Figure 6.17 Control voltages for frequency step, high-gain second-order loop.

Example 6.1:   Limits of the Theory So Far

Assume that a synthesizer is designed with a charge pump, PFD, and loop filter like the ones considered thus far in this chapter. Assume that the loop filter is designed so that the system has a damping constant of 0.707 and a 3-dB bandwidth of 150 kHz. What is the maximum frequency step at the input such that the theory so far is still able to predict the behavior of the system? Provided this condition is met, how long will it take the system to settle from a frequency step?

Solution:
This is a classic question that the authors have been using to torment undergraduate students for years now. Besides being good exam fodder, it also illustrates a very important point.

First, we compute the natural frequency of the loop using approximation #2 from (6.28)

$$\omega_n \approx \frac{\omega_{3dB}}{\left(1 + \zeta\sqrt{2}\right)} = \frac{2\pi \cdot 150 \text{ kHz}}{2} = 2\pi \cdot 75 \text{ kHz}$$

Now referring to Figure 6.16, the maximum normalized phase error to a frequency step is about 0.46 for $\zeta = 0.707$. Therefore, the maximum phase error is

$$\theta_{e\_\max} = 0.46 \frac{\Delta\omega}{\omega_n}$$

The maximum phase error that the PFD can withstand is 2π. Therefore, the largest frequency step that the system can handle is

$$\Delta\omega_{\max} = \frac{\theta_{e\_\max}\omega_n}{0.46} = \frac{2\pi(2\pi \cdot 75\text{kHz})}{0.46} = 6.43 \frac{\text{Mrad}}{s} = 1.02 \text{ MHz}$$

If the frequency step is any larger than this, then the PLL will lose lock and cycle slip, and the transient response will no longer look like Figure 6.16. If it is less than this, Figure 6.16 should do a fair job of predicting the result. In this case it will take a normalized time of $\omega_n t = 7$ before the transient settles or about 14.9 μs.

6.6.2   Nonlinear Transient Behavior

When a PLL is first turned on, or if it experiences a large frequency step at the input, then it may lose lock. In this case, the linear control theory that has been used so far will not apply, as nonlinearities are involved in lock acquisition. The main reason for nonlinearity is the finite linear range of the phase detector. Additionally, there is a finite range over which the loop can acquire lock because VCOs have a finite tuning range. If the loop attempts to lock the VCO to a frequency outside its range, then the loop will never acquire lock. In addition, if the loop has a finite dc gain then the range of lock acquisition may also be limited by the finite range of the phase detector.

In general, a frequency step will result in a nonzero phase error. The general transfer function for phase error is

$$\frac{\theta_e}{\theta_R} = \frac{s}{s + KF(s)} \tag{6.53}$$

Now if a frequency step is applied to this system, the steady-state phase error will be

$$\theta_{e\_ss} = \lim_{s \to 0} \left[\left(\frac{\Delta\omega}{s^2} \right) \cdot \left(\frac{s}{s + KF(s)} \right) \cdot s \right] = \frac{\Delta\omega}{KF(0)} \tag{6.54}$$

Now in the second-order charge pump and PFD system, the steady-state phase error will always be zero because there is an integrator in $F(s)$ and $F(0)$ will go to ∞ in (6.54). In other loops without an integrator in $F(s)$, the phase error would be finite. When the steady-state error exceeds the linear range of the phase detector, the loop will lose lock.

However, in the case of the charge pump and PDF loop, this is not an issue, as in lock, the steady-state phase error is always zero. In this case, the locking range is determined exclusively by the VCO.

So if the loop is going to experience a transient frequency step, then how long does it take the loop to reacquire lock? In this situation, the loop goes into a frequency acquisition mode and the output of the PFD (in frequency detection mode) will look something like that shown in Figure 6.7. In this case, the charge pump will put out pulses of current of value I, which vary in width between almost a complete reference period and a width of almost zero. Therefore, the average current produced by the charge pump until the loop acquires lock will be approximately $I/2$. If it is assumed that all of this current flows onto the capacitor C_1, then the change in voltage across the capacitor as a function of time will be:

$$\frac{\Delta v_C}{\Delta t} = \frac{I}{2C_1} \tag{6.55}$$

Therefore, the settling time will be

$$T_s = \frac{2\Delta v_C C_1}{I} \tag{6.56}$$

From this equation, and making use of the relationship in (6.4) and (6.20), the settling time can be determined for an input frequency change $\Delta\omega$ as

$$T_s = \frac{2C_1 \Delta\omega N}{I K_{VCO}} = \frac{\Delta\omega}{\pi\omega_n^2} \tag{6.57}$$

It should be noted that Δv_C and K_{VCO} relate to the change in VCO frequency, not to the change of input frequency. This leads to a factor of N in the equation. Thus, one can see directly that as the loop bandwidth is expanded, the settling time will decrease as expected. Even though this is the main result in which we are interested, a few more details of the transient behavior of the control voltage are very interesting to examine. To start this discussion, we will assume that the loop filter does not have a capacitor C_2 and is currently charging up towards lock as shown in Figure 6.18.

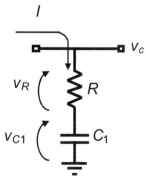

Figure 6.18 Simplified loop filter to illustrate settling behavior.

In this case, the charge pump current will alternately turn on and turn off. When the charge pump current is off, then v_R will be zero and the control voltage v_c will be equal to the voltage across the capacitor v_{C1}. However, when the charge pump current is on, then v_c will be equal to $v_c = v_{c1} + IR$, where IR is the voltage drop across the resistor. This is illustrated in Figure 6.19.

When the capacitor C_2 is included, its filtering effect results in the behavior being a little more complicated. In this case, when the charge pump is on, most of the current still flows into C_1, which still happily charges towards lock, but when it turns off there is no longer an instantaneous change in v_c. In this case, C_2 keeps v_c high, and current flows from C_2 back into C_1 through R. This is shown in Figure 6.20.

Thus, the presence of C_2 tends to smooth out the ripple on the control voltage. For context, the same plot as the one in Figure 6.19 is shown in Figure 6.21(b) where voltage waveforms are also plotted over a larger percentage of acquisition in Figure 6.21(a).

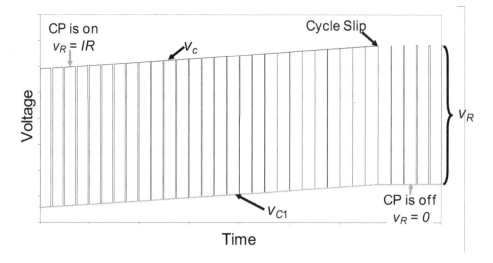

Figure 6.19 Example showing the voltages on the loop filter during acquisition (no C_2 present).

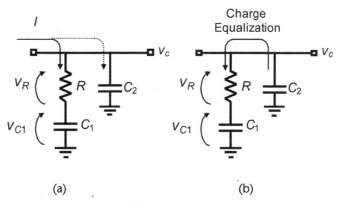

(a)   (b)

Figure 6.20   Illustration of loop filter behavior (C_2 present): (a) charge pump is on and charging both C_1 and C_2; and (b) charge pump is off, and C_2 is discharging into C_1.

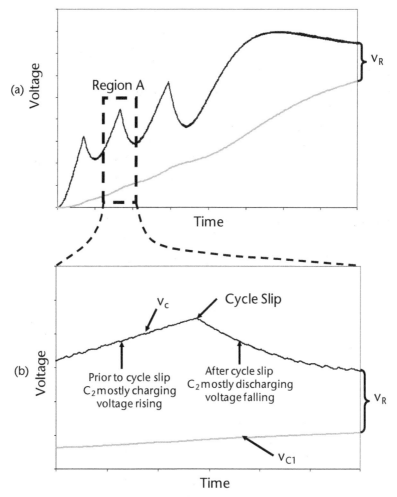

Figure 6.21   Example showing the voltages on the loop filter during acquisition (C_2 present): (a) complete settling and (b) zoom in on region A.

Example 6.2:   Simulation and Estimation of Loop Settling Times

A 3.7–4.3-GHz synthesizer with a step size of 1 MHz is required. A 40-MHz crystal oscillator, a charge pump with $2\pi\cdot100$ μA output current, and a VCO (operating from a 3-V supply) are available. Design a fractional N synthesizer with a loop bandwidth of 150 kHz using these components. Estimate the settling time of the loop for a 30 MHz and a 300-MHz frequency step. Simulate and compare.

Solution:

First, if the VCO is operating with a 3-V supply and must have a 600-MHz tuning range, we can estimate that its K_{vco} will be 200 MHz/V. In addition, as we know the charge pump current, we know that that the K_{phase} will be 100 $\mu A/rad$. For a VCO with nominal frequency of 4 GHz and a reference frequency of 40 MHz, the division ratio will be 100. The next step is to size the loop filter. A 3-dB frequency of 150 kHz requires a natural frequency of 75 kHz (see Example 6.1). Thus, components can be determined as

$$C_1 = \frac{IK_{VCO}}{2\pi\cdot N\omega_n^2} = \frac{2\pi\cdot100\mu A\cdot\left(2\pi\cdot200\,\dfrac{MHz}{V}\right)}{2\pi\cdot100(2\pi\cdot75kHz)^2} = 5.66\,nF$$

To set R, we need to pick a damping constant. Let us pick $1/\sqrt{2}$, or 0.707, which is a popular choice. Now

$$R = 2\zeta\sqrt{\frac{2\pi\cdot N}{IK_{VCO}C_1}} = 2\left(\frac{1}{\sqrt{2}}\right)\sqrt{\frac{2\pi\cdot100}{2\pi\cdot100\mu A\left(2\pi\cdot200\,\dfrac{MHz}{V}\right)\cdot5.66nF}} = 530\Omega$$

and we will set $C_2 = 566$ pF at one-tenth the value of C_1.

Now, a step in output frequency of 30 MHz and 300 MHz corresponds to a step in the reference frequency of 0.3 MHz and 3 MHz, respectively. We learned in Example 6.1 that the maximum input frequency step that can be tolerated for a system with these parameters is 1 MHz. Therefore, the first frequency step will be a linear one and the output will follow the theory of the previous section. Therefore, we expect that it will take approximately 15 μs to settle, as we discovered previously.

In contrast, the second frequency step will involve cycle slipping. For this non-linear case, we use the formula given in (6.57) to estimate the acquisition time as:

$$T_s = \frac{\Delta\omega}{\pi\omega_n^2} = \frac{2\pi\cdot3MHz}{\pi(2\pi\cdot75kHz)^2} = 27\mu s$$

Therefore, complete settling in this case should take 27 μs plus an additional 15 μs for phase lock.

This behavior can be simulated using ideal components in a simulator such as Cadence's Spectre. The blocks for the divider, VCO, PFD, and charge pump can be programmed using ideal behavioral models. These can be connected to the loop

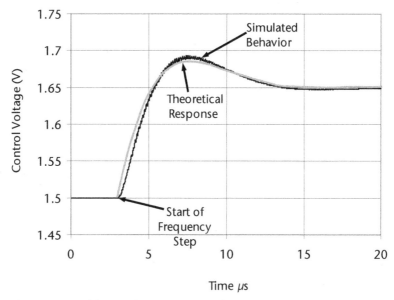

Figure 6.22 Response of the PLL design's control voltage during a 30-MHz frequency step.

filter that has just been designed. From these simulations, we can look at the control voltage on the VCO to verify the performance of the loop. A plot of the response of the system to a 0.3 MHz step at its input, compared to simple theory is shown in Figure 6.22. From this figure, it is easy to see that the simple theory does an excellent job of predicting the settling behavior of the loop with only a slight deviation. This small discrepancy is most likely due to the presence of C_2 and to the sampling nature of the loop components.

The second frequency step can be simulated as well. The results of this simulation are plotted in Figure 6.23 and compared to the linear voltage ramp suggested by simple theory previously. In this case, this plot shows that the nonlinear response

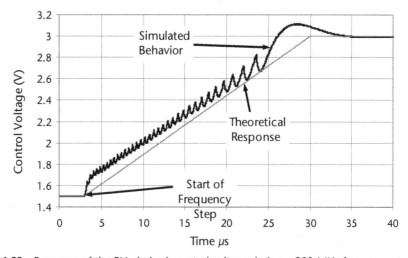

Figure 6.23 Response of the PLL design's control voltage during a 300-MHz frequency step.

is predicted fairly well by the simple formula; however, the actual response is slightly faster. The tail of this graph is cut off, but the loop settled in about 39 μs, which is very close to the 42 μs predicted. The main difference between the simple estimate and reality is the fact that phase acquisition begins before the PLL actually reaches its final frequency value. We predicted it would begin in this simulation at 30 μs (when the simple theory predicts that the voltage ramp will reach its final value), but the linear portion of the graph actually starts earlier than this at about 25–26 μs. This accounts for our slightly pessimistic estimate. Still, such a simple estimate is remarkably good at predicting quite complicated behavior.

6.6.3   Various Noise Sources in PLL Synthesizers

6.6.3.1   VCO Noise

All the circuits in the synthesizer contribute to the overall noise in different ways and the noise they produce has different characteristics. For instance, the phase noise from a VCO can be described as [60]

$$\varphi^2 \text{vco}(\Delta\omega) = \frac{C}{\Delta\omega^2} + D \qquad (6.58)$$

where C is a constant of proportionality and $\Delta\omega$ is the frequency offset from the carrier. Thus, at most frequencies of interest, the phase noise produced by the VCO will go down at 20 dB/decade as we move away from the carrier. This will not continue indefinitely, as thermal noise will put a lower limit on this phase noise D which, for most integrated VCOs, is somewhere between –120 and –150 dBc/Hz. VCO phase noise is usually dominant outside the loop bandwidth and of less importance at low offset frequencies.

6.6.3.2   Crystal Reference Noise

Crystal resonators are widely used in frequency control applications because of their unequaled combination of high Q, stability and small size. The resonators are classified according to "cut," which is the orientation of the crystal wafer (usually made from quartz) with respect to the crystallographic axes of the material. The total noise power spectral density of a crystal oscillator can be found by Leeson's formula [64]:

$$\varphi^2_{XTAL}(\Delta\omega) = 10^{-16\pm1} \cdot \left[1 + \left(\frac{\omega_0}{2\Delta\omega \cdot Q_L} \right)^2 \right] \left[1 + \frac{\omega_c}{\Delta\omega} \right] \qquad (6.59)$$

where ω_0 is the oscillator output frequency, ω_c is the corner frequency between $1/f$ and thermal noise regions, which is normally in the range 1~10 kHz, and Q_L is the loaded quality factor of the resonator. Because Q_L for a crystal resonator is very large (normally in the order of 10^4 to 10^6), the reference noise contributes only to the very close-in noise and it quickly reaches the thermal noise floor at an offset frequency around ω_c.

6.6.3.3 Frequency Divider Noise

Frequency dividers consist of switching logic circuits, which are sensitive to clock timing jitter. The jitter in the time domain can be converted to phase noise in the frequency domain. Time jitter or phase noise occurs when rising and falling edges of digital dividers are superimposed with spurious signals such as thermal noise and flicker noise in semiconductor materials. Ambient effects result in variation of the triggering level due to temperature and humidity. Frequency dividers generate spurious noise especially for high-frequency operation. While there is no substitute for real simulations or measurements of a particular divider, Kroupa provided an empirical formula that describes the amount of phase noise that typical frequency dividers add to a signal [16, 17]

$$\varphi^2_{\text{Div\_Added}}(\Delta\omega) \approx \frac{10^{-14\pm1} + 10^{-27\pm1}\,\omega^2_{\text{do}}}{2\pi\cdot\Delta\omega} + 10^{-16\pm1} + \frac{10^{-22\pm1}\,\omega_{do}}{2\pi} \tag{6.60}$$

where ω_{do} is the divider output frequency and $\Delta\omega$ is the offset frequency. Notice that the first term in (6.60) represents the flicker noise and the second term gives the white thermal noise floor. The third term is caused by timing jitter due to coupling, ambient and supply variations. Additional information may be found in [2].

6.6.3.4 Phase Detector Noise

Phase detectors experience both flicker and thermal noise. Although the noise generated by phase detectors depends on a number of factors such as circuit topology and the technology used to implement them, a rule of thumb is that at large offsets, phase detectors generate a white phase noise floor of about –160 dBc/Hz, which is thermal noise dominant. The noise power spectral density of phase detectors is given by [55]

$$\varphi^2_{\text{PD}}(\Delta\omega) \approx \frac{2\pi\cdot10^{-14\pm1}}{\Delta\omega} + 10^{-16\pm1} \tag{6.61}$$

Additional information may be found in [2].

6.6.3.5 Charge Pump Noise

The noise of the charge pump can be characterized as an output noise current and is usually given in pA/$\sqrt{\text{Hz}}$. Note that at this point in the loop the phase is represented by the current. The charge pump output current noise can be a strong function of the frequency and of the width of the current pulses, therefore for low noise operation it is desirable to keep the charge pump currents matched as well as possible. This is because current sources only produce noise when they are on. In an ideal loop, when locked, the charge pump is always off. However, nonidealities result in finite pulses, but the closer reality matches the ideal case, the less noise will be produced. Also, note that as the frequency is decreased, 1/f noise will become more important, causing the noise to increase. This noise can often be the dominant noise source at low-frequency offsets. Charge pump noise can be simulated with proper tools and the results depend on the design in question so no simple formula will be given here.

6.6.3.6 Loop Filter Noise

Loop filters are simple RC circuits and can be analyzed for noise in the frequency domain in a linear manner. The most common loop filter that has been examined in this chapter will now be analyzed. It contains only one noise source, the thermal noise associated with R. Thus, the loop filter with associated noise can be drawn as shown in Figure 6.24. Now the noise voltage develops a current flowing through the series combination of C_1, C_2, and R (assuming that the charge pump and VCO are both open circuits), which is given by

$$i_{n\_\text{LPF}} = \frac{1}{R} \cdot \frac{v_n s}{s + \dfrac{C_1 + C_2}{C_1 C_2 R}} \approx \frac{1}{R} \cdot \frac{v_n s}{s + \dfrac{1}{C_2 R}} \tag{6.62}$$

Thus, this noise current will have a highpass characteristic, and therefore the loop will not produce any noise at dc and this noise will increase with frequency until the highpass corner is reached, after which it will be flat.

6.6.3.7 ΣΔ Noise

If a ΣΔ noise shaper is used to control the divide ratio in a synthesizer, then it will also generate phase noise. Fortunately, discrete fractional spurs become more like random noise after ΣΔ noise shaping which is the main reason these more complicated circuits are preferred over the simple accumulators discussed later in this chapter. The single-sideband phase noise of the ΣΔ modulator is given by

$$\frac{\varphi_{\Sigma\Delta}^2(f)\left[rad^2/Hz\right]}{2} = \frac{(2\pi)^2}{24 f_r} \cdot \left[2\sin\left(\frac{\pi f}{f_r}\right)\right]^{2(m-1)}$$

$$PN_{\Sigma\Delta}(f)\left[dBc/Hz\right] = 10\log\left\{\frac{(2\pi)^2}{24 f_r} \cdot \left[2\sin\left(\frac{\pi f}{f_r}\right)\right]^{2(m-1)}\right\} \tag{6.63}$$

where f is the offset frequency, f_r is the reference sampling frequency, and m is the order of the ΣΔ modulator. Note that the noise is shaped by the modulator. The higher the order of the modulator, the more noise is pushed to higher frequencies where it can be filtered by the PLL.

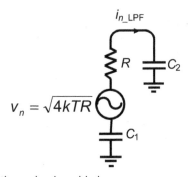

Figure 6.24 Loop filter with thermal noise added.

6.6.4 In-Band and Out-of-Band Phase Noise in PLL Synthesis

The noise transfer functions for the various noise sources in the loop can be derived quite easily using the theory already presented. There are two noise-transfer functions: one for the VCO and one for all other sources of noise in the loop. All noise generated by the PFD, charge pump, divider, and loop filter is referred back to the input, and the noise from the VCO is referred to the output as shown in Figure 6.25. The transfer function for $\varphi_{\text{noiseI}}(s)$ has already been derived (as it is the same as the loop transfer function in the continuous domain) and is given by

$$\frac{\varphi_{\text{noise out}}(s)}{\varphi_{\text{noiseI}}(s)} = \frac{F(s)K_{\text{VCO}}K_{\text{phase}}}{s + \dfrac{F(s)K_{\text{VCO}}K_{\text{phase}}}{N}} \tag{6.64}$$

where for the charge pump PFD loop, the transfer function for the filter, divider, and crystal reference noise becomes

$$\frac{\varphi_{\text{noise out}}(s)}{\varphi_{\text{noiseI}}(s)} = \frac{\dfrac{IK_{\text{VCO}}}{2\pi \cdot C_1}(1 + RC_1 s)}{s^2 + \dfrac{IK_{\text{VCO}}}{2\pi \cdot N}Rs + \dfrac{IK_{\text{VCO}}}{2\pi \cdot NC_1}} \tag{6.65}$$

This function is a lowpass filter. This means that for low frequencies (inside the loop bandwidth), the loop will track the input phase (which includes the phase noise) and thus this noise will be transferred to the output. At higher offset frequencies, (outside the loop bandwidth) this noise is suppressed as the loop prevents the VCO from following these changes in phase. Also note that the division ratio plays a very important part in this transfer function. It can be seen that at low reference frequencies, where the s and s^2 terms in (6.65) can be ignored relative to the constant terms, higher division ratio N directly results in higher phase noise. This is one of the strongest arguments for using fractional-N architectures in synthesizers, as with large division ratios it is hard to get low phase noise performance.

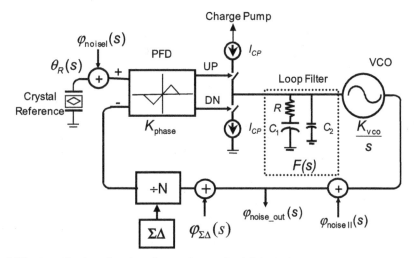

Figure 6.25 A synthesizer showing places where noise is injected.

The transfer function for the noise due to the VCO is slightly different. In this case, input noise is set to zero and then the transfer function is derived in the usual way. It is given in general by

$$\frac{\varphi_{\text{noise out}}(s)}{\varphi_{\text{noise II}}(s)} = \frac{s}{s + \dfrac{F(s)K_{\text{VCO}}K_{\text{phase}}}{N}} \tag{6.66}$$

Using our loop will give

$$\frac{\varphi_{\text{noise out}}(s)}{\varphi_{\text{noise II}}(s)} = \frac{s^2}{s^2 + \dfrac{IK_{\text{VCO}}}{2\pi \cdot N}Rs + \dfrac{IK_{\text{VCO}}}{2\pi \cdot NC_1}} \tag{6.67}$$

This is a highpass filter. Thus, at low offsets inside the loop bandwidth, the VCO noise is suppressed by the feedback action of the loop, but outside the loop bandwidth, the VCO is essentially free running and thus the loop noise approaches the VCO noise.

This noise from the $\Sigma\Delta$ is injected into the system in a third location as shown in Figure 6.25. Therefore, its noise transfer function to the output is given by

$$\frac{\varphi_{\text{noise\_out}}(s)}{\varphi_{\Sigma\Delta}(s)} = \frac{\dfrac{F(s)K_{\text{VCO}}K_{\text{phase}}}{N}}{s + \dfrac{F(s)K_{\text{VCO}}K_{\text{phase}}}{N}} \tag{6.68}$$

Note that, due to the highpass nature of the $\Sigma\Delta$ noise transfer function, the order of the loop rolloff is very important. It can be seen from (6.63) that the noise of a $\Sigma\Delta$ modulator has a noise shaping slope of $20(m-1)$ dB/decade, while an nth-order lowpass loop filter has a roll-off slope of $20n$ dB/decade. Therefore, the order of the loop filter must be higher than or equal to the order of the $\Sigma\Delta$ modulator in order to attenuate the out-of-band noise due to $\Sigma\Delta$ modulation. Thus, for instance, when calculating the effect of the $\Sigma\Delta$ modulator on out-of-band noise, it is necessary to include the effect of C_2 in the loop filter, as this will provide extra attenuation out of band. In this case the $\Sigma\Delta$ noise transfer function to the output would be

$$\frac{\varphi_{\text{noise\_out}}(s)}{\varphi_{\Sigma\Delta}(s)} = \frac{K_{\text{VCO}}K_{\text{phase}}(1 + sC_1R)}{s^2N(C_1 + C_2)(1 + sC_sR) + K_{\text{VCO}}K_{\text{phase}}(1 + sC_1R)} \tag{6.69}$$

where

$$C_s = \frac{C_1C_2}{C_1 + C_2}.$$

Example 6.3:   System Phase Noise Calculations

Estimate the phase noise for the synthesizer that was designed in Example 6.2. The VCO has a phase noise of –120 dBc/Hz at 1-MHz offset (it bottoms out at –130 dBc/Hz), and the charge pump puts out a noise current of 10pA/$\sqrt{\text{Hz}}$. Ignoring PFD, divider and reference noise sources plot the phase noise. In addition, what would the phase noise of an equivalent integer-N design be?

Solution:

From Example 6.2, we know the charge pump current and we know that that the K_{phase} will be 100 μA/rad. Now in the case of the integer-N synthesizer the reference must be 1 MHz to get a step size of 1 MHz. Therefore, the division ratio will be 4,000. Knowing that we want a loop bandwidth of 150 kHz means that we need a natural frequency of 75 kHz (assuming a damping constant of 0.707) and this means that for the integer-N design, C_1 and R are 141.5 pF and 21.2 kΩ, respectively. Now we will assume that the VCO follows the 20 dB/decade rule just outlined. Therefore, we can come up with a linear expression for the phase noise of the VCO based on (6.58):

$$C = \log^{-1}\left(\frac{PN_{VCO}}{10}\right) \cdot \Delta\omega^2 = \log^{-1}\left(\frac{-120}{10}\right) \cdot \left(2\pi \cdot 1\text{MHz}\right)^2 = 39.5 \frac{\text{rad}^4}{\text{Hz}^2}$$

Now, as the VCO bottoms out at –130 dBc/Hz, that means we can determine the D term of the VCO phase noise equation (6.58):

$$D = \log^{-1}\left(\frac{PN_{VCO}}{10}\right) = \log^{-1}\left(\frac{-130}{10}\right) = 10^{-13} \frac{\text{rad}^2}{\text{Hz}}$$

This can be turned into an equation that has units of voltage instead of units of V^2:

$$\varphi_{VCO}\left(\Delta\omega\right) = \sqrt{\frac{39.5}{\Delta\omega^2} + 10^{-13}} \frac{\text{rad}}{\sqrt{\text{Hz}}}$$

The output noise current from the charge pump can be input referred by dividing by K_{phase}:

$$\text{Noise}_{CP} = \frac{i_n}{K_{phase}} = \frac{10\frac{pA}{\sqrt{\text{Hz}}}}{100\frac{\mu A}{\text{rad}}} = 100n \cdot \frac{\text{rad}}{\sqrt{\text{Hz}}}$$

The noise from the loop filter must also be moved back to the input:

$$\text{Noise}_{LFP}(\omega) = \frac{1}{K_{phase}}\left|\frac{\sqrt{\frac{4kT}{R}}j\omega}{j\omega + \frac{1}{C_2 R}}\right|$$

Clearly, noise from the lowpass filter is dependent on filter component values as well as the phase detector gain. Now input-referred noise from the loop filter and the charge pump can each be substituted into (6.65) while noise from the VCO can be substituted into (6.67) to determine the contribution to the phase noise at the output. Once each component value at the output is calculated, the overall noise can be computed (noting that noise adds as power). So, for instance, in the case of the noise due to the charge pump, the output phase noise for the fractional-N case is [from (6.65)]

$$\varphi_{\text{noise out\_CP}}(s) = \frac{2.22 \cdot 10^{13} \left(1 + 3 \cdot 10^{-6} s\right)}{s^2 + 6.66 \cdot 10^5 s + 2.22 \cdot 10^{11}} 100n \cdot \frac{\text{rad}}{\sqrt{\text{Hz}}}$$

Therefore, to plot phase noise in dBc/Hz, we take

$$PN_{\text{CP}}(\Delta\omega) = 20 \log\left(\left|\frac{2.22 \cdot 10^{13} \left(1 + 3 \cdot 10^{-6} j\Delta\omega\right)}{(j\Delta\omega)^2 + 6.66 \cdot 10^5 j\Delta\omega + 2.22 \cdot 10^{11}}\right| 100n \cdot \frac{\text{rad}}{\sqrt{\text{Hz}}}\right)$$

The results of this calculation and similar ones for the other noise sources are shown in Figure 6.26. The total phase noise is computed by

$$\varphi_{\text{total}} = \sqrt{\varphi_{\text{noise\_out\_CP}}^2 + \varphi_{\text{noise\_out\_VCO}}^2 + \varphi_{\text{noise\_out\_LPF}}^2}$$

and is shown in the figure.

If we assume that the numbers given so far have been for single-sideband phase noise, then we can also compute the integrated phase noise of this design as:

$$\text{IntPN}_{\text{rms}} = \frac{180\sqrt{2}}{\pi} \sqrt{\int_{f=10\text{kHz}}^{f=10\text{MHz}} \varphi_{\text{total}}^2 df} = 0.41°$$

Note, in this example, that the loop-filter noise is quite low, and could have been ignored safely. Also note that due to the frequency response of the filter even in-band, noise from the loop filter is attenuated at lower frequencies. Inside of the

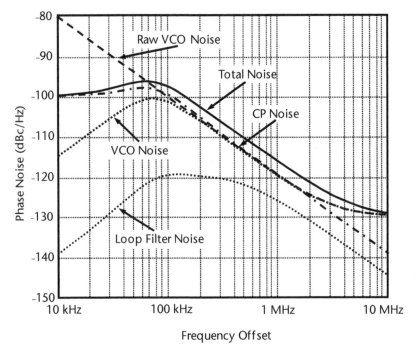

Figure 6.26 Phase noise of various components and overall phase noise for the system with fractional-*N* divider.

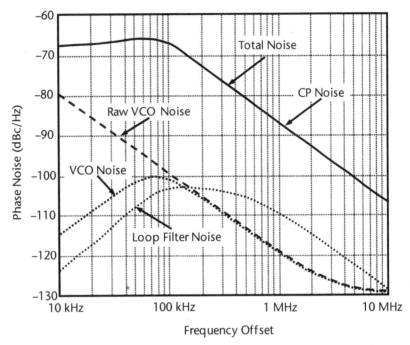

Figure 6.27 Phase noise of various components and overall phase noise for the system with integer-*N* divider.

loop bandwidth, the total noise is dominated by the charge pump, which is the more dominant in-band noise source. Note that out of band, the noise is slightly higher than the VCO noise. This is because the charge pump is still contributing. This could be corrected by making the loop bandwidth slightly smaller and thus attenuating the out-of-band contribution of the charge pump by a few more decibels.

With the integer-*N* numbers, the phase noise is shown in Figure 6.27. Note that with integer *N*, the noise is completely dominated by the charge pump, both inside and outside of the loop bandwidth. To reduce the effect of charge pump noise, the loop bandwidth in this case should be reduced by at least two orders of magnitude. Note also the dramatic change in the in-band phase-noise performance between the two designs. While the fractional design has −100 dBc/Hz of in-band noise, this design has a performance of only −67 dBc/Hz. Note that the two numbers are different by 20log(40), which is the ratio of the two divider values, as would be expected.

6.7 Reference Feedthrough

Reference feedthrough can affect how much time the charge pump must remain on in the locked state. If the phase difference between the reference and the feedback from the VCO is zero, then both the current sources are on for an instant while the PFD resets itself. If the two currents were mismatched in that instant, then a net charge would be deposited onto the loop filter, forcing a correction during the next cycle of the reference. This will create an ac signal on the control line of the VCO at the reference frequency, as shown in Figure 6.28. If the current sources are mismatched by an amount ΔI and the reset path has a delay of time δ, then a charge q is placed on the loop filter, where

$$q = \delta \cdot \Delta I \tag{6.70}$$

This will require the other current source to be on for a time t given by

$$t = \frac{\delta \cdot \Delta I}{I_{CP}} \tag{6.71}$$

to remove the charge. Thus, the total time that the CP will be on over one cycle will be

$$t_{CP} = \delta \left(1 + \frac{\Delta I}{I_{CP}} \right) \tag{6.72}$$

Now, as these current pulses happen at the reference frequency and well beyond the corner frequency of the loop, and if a typical loop filter is assumed, then all this current will flow into C_2 of Figure 6.28. Thus, the voltage will have a triangular shape and will have a peak amplitude of

$$V_{mm} = \frac{\delta \cdot \Delta I}{C_2} \tag{6.73}$$

From basic Fourier series analysis, this triangle wave will have a fundamental component with an amplitude of

$$V_{ref} = \frac{\delta^2 \cdot \Delta I}{C_2 T} \left(1 + \frac{\Delta I}{I_{CP}} \right) \tag{6.74}$$

provided that the current flows for a small fraction of a period T. This signal on the control line of the VCO will modulate the output of the VCO according to the formula

$$v_{out}(t) \approx A\cos(\omega_o t) + \frac{AV_{ref}K_{VCO}}{2\omega_{ref}} \left[\cos(\omega_o + \omega_{ref})t - \cos(\omega_o - \omega_{ref})t \right] \tag{6.75}$$

Thus, the magnitude of the reference spurs will be

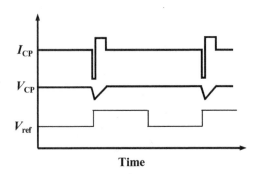

Figure 6.28 Graph showing the effect of mismatched current in a CP.

$$\text{Spurs} = 20\log\left[\frac{\delta^2 \cdot \Delta I \cdot K_{\text{VCO}}}{4\pi \cdot C_2}\left(1 + \frac{\Delta I}{I_{\text{CP}}}\right)\right] \tag{6.76}$$

dBc below the carrier.

Example 6.4:   Reference Feedthrough Due to CP Mismatch

Starting with the loop in Example 6.2, if the UP current source were 10% high, what reference feedthrough would result if the reset delay in the PFD/CP is assumed to be 5 ns?

Solution:
In Example 6.2, values of note were $C_2 = 566$ pF, $I_{\text{CP}} = 628$ μA, and $K_{\text{VCO}} = 200$ MHz/V. In this case the UP current is assumed to be 691 μA and therefore $\Delta I = 63$ μA. The spurs can be predicted from (6.76)

$$\text{Spurs} = 20\log\left[\frac{\delta^2 \cdot \Delta I \cdot K_{\text{VCO}}}{4\pi \cdot C_2}\left(1 + \frac{\Delta I}{I_{\text{CP}}}\right)\right] =$$

$$20\log\left[\frac{(5\text{ns})^2 \cdot 63\mu A \cdot \left(2\pi \cdot 200\frac{\text{MHz}}{\text{V}}\right)}{4\pi \cdot 566 \text{ pF}}\right]\left(1 + \frac{63\ \mu A}{628\ \mu A}\right) = -71\ \text{dBc}$$

This can also be simulated. A result of an FFT of the output of the synthesizer with a 10% mismatch in CP currents is shown in Figure 6.29. In this plot, simulations show that the spurs are at a level of –69.5 dBc, which is very close to what was predicted from the theory above.

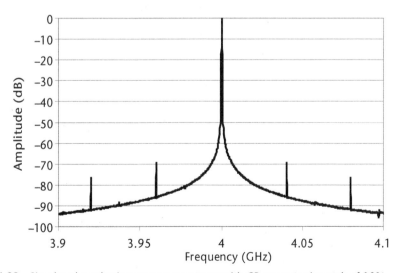

Figure 6.29 Simulated synthesizer output spectrum with CP current mismatch of 10%.

6.8   Fractional-*N* Frequency Synthesizers

In contrast to an integer-N synthesizer, a fractional-N synthesizer allows the PLL to operate with high reference frequency while achieving a fine step size by constantly swapping the loop division ratio between integer numbers. As a result, the average division ratio is a fractional number. As will be shown, a higher reference frequency leads to lower in-band phase noise and faster PLL transient response. In addition, for multiband applications, often the channel spacing of the different bands is skewed, requiring an even lower reference frequency if the synthesizer is to cover both bands.

Example 6.5:   The Problem with Using an Integer-*N* Synthesizer for Multiband Applications

Determine the maximum reference frequency of an integer-N frequency synthesizer required to cover channels from 2,400 MHz to 2,499 MHz spaced 3 MHz apart, and channels from 5,100 MHz to 5,200 MHz spaced 4 MHz apart.

Solution:
If an integer-N synthesizer were designed to service only one of these bands, then it would have a maximum reference frequency of 3 MHz in the first case and 4 MHz in the second case. However, if a synthesizer must be designed to cover both of these bands, then its step size must be 1 MHz to allow it to tune exactly to every frequency required.

    In the simplest case, the fractional-N synthesizer generates a dynamic control signal that controls the divider, changing it between two integer numbers. By toggling between the two integer division ratios, a fractional division ratio can be achieved by time-averaging the divider output. As an example, if the control changes the division ratio between 8 and 9, and the divider divides by 8 for 9 divider output cycles and by 9 for 1 divider output cycle and then the process repeats itself, then the average division ratio will be

$$\bar{N} = \frac{\text{Total Divider Input Pulses Needed}}{\text{Total Divider Output Pulses Generated}} = \frac{8 \times 9 + 9 \times 1}{10} = 8.1 \qquad (6.77)$$

If the divider were set only to divide by 8, then it would produce 10 output pulses for 80 input pulses. However, now it will take 81 input pulses to produce 10 output pulses. Thus, in other words, the device swallows one extra input pulse to produce every 10 output pulses. In the PLL synthesizer, this time average is dealt with by the transfer function in the loop. This transfer function will always have a lowpass characteristic. Thus, it will deliver the average error signal to the VCO. As a result, the output frequency will be the reference frequency multiplied by the average division ratio. However, toggling the divider ratio between two values in a repeating manner generates a repeating time sequence. In the frequency domain, this periodic sequence will generate spurious components (or spurs) at integer multiples of the repetition rate of the time sequence. Such spurious components can be reduced by using $\Sigma\Delta$ modulators in which the instantaneous division ratio is randomized while maintaining the correct average value.

6.8.1   Fractional-*N* Synthesizer with Dual Modulus Prescaler

Figure 6.30 illustrates one way to implement a simple fractional-*N* frequency synthesizer with a dual-modulus prescaler $P/P + 1$. Note that it is called a dual modulus prescaler because it can be programmed to two division ratios. As discussed in the previous section, the fractionality can be achieved by toggling the divisor value between two values P and $P + 1$. The modulus control signal (C_{out}) is generated using an accumulator (also called an integrator or adder with feedback, or a counter) with size of F (or $\log_2 F$ bits). That is, an overflow occurs whenever the adder output becomes equal to or larger than F. At the *i*th clock rising edge, the accumulator's output y_i can be mathematically expressed as

$$y_i = (y_{i-1} + K_i) \bmod F \qquad (6.78)$$

where y_{i-1} is the output on the previous rising clock edge, and K_i is a user-defined input and its value will determine the fractional divider value. Its use will be illustrated shortly in Example 6.6. Note that the modular operation ($A \bmod B$) returns the remainder of ($A \div B$) and is needed for modeling the accumulator overflow.

Example 6.6:   A Simple Accumulator Simulation

Describe the operation of a 3-bit accumulator with input $K = 1$ and $K = 3$, assuming the accumulator seed value (i.e., the initial accumulator output value) is equal to zero.

Solution:
If the accumulator has 3 bits, the size of the accumulator is $2^3 = 8$, or $F = 8$, even though the largest value that can be stored is 111, corresponding to 7. If input

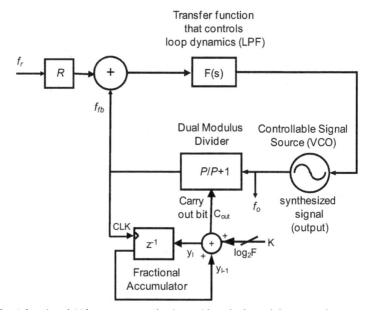

Figure 6.30   A fractional-*N* frequency synthesizer with a dual-modulus prescaler.

Table 6.1 Accumulator Operations with $F = 8$ and $K = 1$

Clock Cycle i	0	1	2	3	4	5	6	7	8	9	10	11	12	13	14	15	16	17	18
y_i	0	1	2	3	4	5	6	7	0	1	2	3	4	5	6	7	0	1	2
y_{i-1}	NA	0	1	2	3	4	5	6	7	0	1	2	3	4	5	6	7	0	1
C_{out}	1	0	0	0	0	0	0	0	1	0	0	0	0	0	0	0	1	0	0

word $K = 1$, namely, 001, the accumulator value y increases by 1 every cycle until it reaches the maximum value that can be represented using 3-bits, namely, $y_{max} = 7 = 111$. After this point, the accumulator will overflow leaving its value $y = 0$ and $C_{out} = 1$. It will take 8 clock cycles for the accumulator to overflow if $K = 1$. In other words, the accumulator size $F = 8$. For $K = 3 = 3\text{'b}011$, the accumulator adds an increment value of 3 every cycle and thus overflows more often. The accumulator value and its carryout C_{out} are summarized cycle by cycle in Table 6.1 for $K = 1$ and Table 6.2 for $K = 3$.

As shown in the above example, for the $K = 1$ case, C_{out} is high for one cycle and low for seven cycles within every eight clock cycles, so the frequency of C_{out} is $f_{clk}/8$. For the $K = 3$ case, C_{out} is high for three cycles and low for five cycles within every eight clock cycles, so the frequency of C_{out} is $3f_{clk}/8$. In general, for a constant input word K, the accumulator carry out will be high for K cycles and will be low for $F - K$ cycles. Also, note that the frequency of C_{out} will be equal to

$$f_{C_{out}} = \frac{K f_{clk}}{F} \tag{6.79}$$

If the dual modulus prescaler divides by P when C_{out} is low and divides by $P + 1$ when C_{out} is high, the average VCO output frequency is

$$f_o = \frac{f_r}{R}\left[\frac{(P+1)K + P(F-K)}{F}\right] = \frac{f_r}{R}\left[P + \frac{K}{F}\right] \tag{6.80}$$

Because fractionality is achieved by using this accumulator, it is often called a fractional accumulator. It has a fixed size F due to a fixed number of accumulator bits built into the hardware. The dual-modulus prescaler ratio P is normally fixed as well. The only programmable parameter for the architecture shown in Figure 6.30 is the accumulator input K, which can be programmed from 1 to a maximum number of F. Thus, because K is an integer, from (6.80) it can be seen that the step size of this architecture is given by

$$\text{Step Size} = \frac{f_r}{RF} \tag{6.81}$$

Table 6.2 Accumulator Operations with $F = 8$ and $K = 3$

Clock Cycle i	0	1	2	3	4	5	6	7	8	9	10	11	12	13	14	15	16	17	18
y_i		0	3	6	1	4	7	2	5	0	3	6	1	4	7	2	5	0	3	6
y_{i-1}		NA	0	3	6	1	4	7	2	5	0	3	6	1	4	7	2	5	0	3
C_{out}		1	0	0	1	0	0	1	0	1	0	0	1	0	0	1	0	1	0	0

where R is normally fixed to avoid changing the comparison frequency at the input. Note that R is normally as small as possible to minimize the in-band phase noise contribution from the crystal. Thus, step size is inversely proportional to the number of bits ($\log_2 F$), so as a result, the accumulator is normally used to reduce synthesizer step size without increasing R and degrading the in-band phase noise. More detail on spur reduction is given in [2].

6.8.2 Fractional-N Synthesizer with Multimodulus Divider

Replacing the dual-modulus divider with a multimodulus divider (MMD), the synthesizer architectures shown in Figure 6.30 can be modified to a more generic form, as illustrated in Figure 6.31. Using a multimodulus divider has the advantage that the range of frequencies over which the synthesizer can be tuned is expanded, compared to the previous architecture. The synthesizer output frequency is given by

$$f_o = \frac{f_r}{R}\left[I + \frac{K}{F}\right] \tag{6.82}$$

where I is the integer portion of the loop divisor and, depending on the complexity of the design, I could have many possible integer values. For instance, if loop division ratio 100.25 is needed, we can program $I = 100$, $K = 1$ and $F = 4$. The MMD division ratio is toggled between 100 and 101.

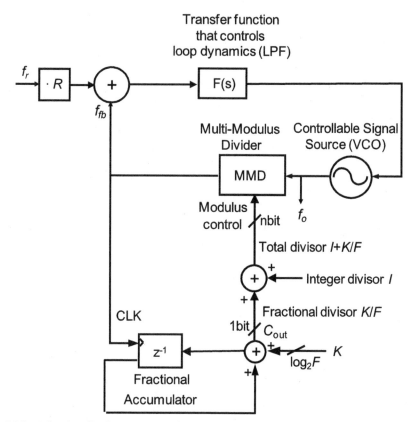

Figure 6.31 A fractional-N frequency synthesizer with a multimodulus divider.

In a popular MMD topology using cascaded 2/3 cells [53] with an n-bit modulus control signal, the MMD division ratio is given by:

$$N_{MMD} = P_1 + 2^1 P_2 + \ldots 2^{n-2} P_{n-1} + 2^{n-1} P_n + 2' \qquad (6.83)$$

where the MMD programming range is 2^n to $2^{n+1} - 1$. For instance, a 6-bit MMD can be programmed from 64 to 127. The MMD program range can be further extended through the use of additional logic. Wide programming range is critical for multiband frequency synthesis, especially when sigma-delta noise shaping is employed. More details of divider design are given in [2].

6.8.3 Fractional-N Spurious Components

The above-discussed fractional-N architectures suffer from a common side effect of generating spurious components associated with periodically toggling the loop division ratio. Recall that any repeatable pattern in the time domain causes spurious tones in the frequency domain. The fractional accumulator periodically generates the carry out that toggles the loop division ratio. It is expected that there should be spurious tones at multiples of the carryout frequency $(f_r/R)\cdot(K/F)$. In the following example, fractional-N spurs are analyzed with simulation and measurement.

Example 6.7:   The Use of Accumulators in Fractional-N Synthesizers

Design a fractional-N synthesizer architecture for synthesizing 11 channels from 819.2 MHz to 820.96 MHz with a step size of 160 kHz and reference comparison frequency of $f_r/R = 5.12$ MHz. Determine the frequencies of fractional-N spurious components.

Solution:
The synthesizer step size is given by $\dfrac{f_r}{R} \cdot \dfrac{1}{F} = 160$ kHz. Because the comparison frequency is $f_r/R = 5.12$ MHz, the fractional accumulator size can be chosen as

$$F = \frac{f_r}{R} \cdot \frac{1}{160\text{ kHz}} = \frac{5120\text{ kHz}}{160\text{ kHz}} = 32$$

which can be implemented using a 5-bit accumulator. The accumulator input, that is, the fine tune frequency word K, can be programmed from 0 to 10 to cover the 11 channels from 819.2 MHz to 820.96 MHz with step size of 160 kHz (the first channel does not require any fractionality). The integer divisor ratio, that is, the coarse tune frequency word I, can be determined by the channel frequency. For instance, the first channel frequency is synthesized as

$$\frac{f_r}{R}\left(I + \frac{0}{F}\right) = \frac{f_r}{R} \cdot I = 819.2\text{ MHz}$$

which leads to $I = 160$. Hence, the loop total divisor is given by $N = 160 + K/32$, where $K = 0, 1, \ldots, 10$. If a dual-modulus divider is used to construct a fractional-N synthesizer, as illustrated in Figure 6.30, the dual-modulus divider ratio should be

$P/P + 1 = 160/161$, which is not the best solution as far as the power consumption and circuit speed is concerned. There are better circuit implementations, such as using a multimodulus divider or a pulse-swallow divider, which allows much of the implemented circuitry to operate at much lower speeds. If a fractional-N architecture with MMD is used, as illustrated in Figure 6.31, a 7-bit MMD with programmable range from 128 to 255 is needed based on (6.83). In any of the above solutions, the loop divisor of the fractional-N architecture is toggled between 160 and 161. The modulus control is generated by the accumulator carryout, which has a frequency of $(f_r/R)\cdot(K/F)$. Thus, the loop is divided by 160 for K reference cycles and divided by 161 for $F - K$ cycles, which results in an average division ratio of $160 + K/F$. As an example, for the second channel with $K = 1$ and $F = 32$, the simulated fractional accumulator output is given in Figure 6.32. As shown, the fractional accumulator outputs a carryout in every 32 comparison cycles, which forces the loop divider to divide by 160 for 31 cycles and to divide by 161 for 1 cycle [Figure 6.32(a)]. The periodic phase correction pulse due to dividing by 161 generates fractional spurs with uniform spacing of $f_r/R/32 = 160$ kHz as shown in Figure 6.32(b).

The spurious tones generated by the accumulator will appear in the output spectrum of the synthesizer. Figure 6.33 presents a measured spectrum at a fractional-N synthesizer output with loop divisor $N = 160 + 1/32$ and the comparison frequency $f_r/R = 5.12$ MHz. Fractional spurs at integer multiples of $\dfrac{f_r}{R}\cdot\dfrac{K}{F} = 160$ kHz are observed. The roll-off of the spur magnitude as frequency increases is due to the loop lowpass roll-off characteristics.

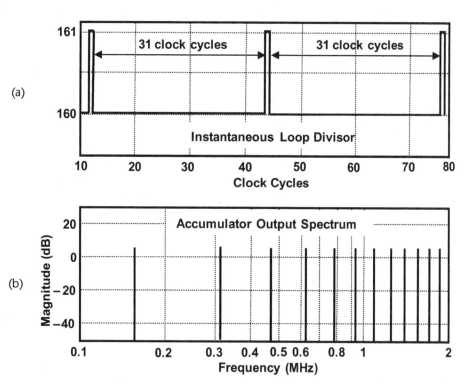

Figure 6.32 Simulated fractional accumulator output with loop divisor $N = 160+1/32$ and the comparison frequency $f_r/R = 5.12$ MHz.

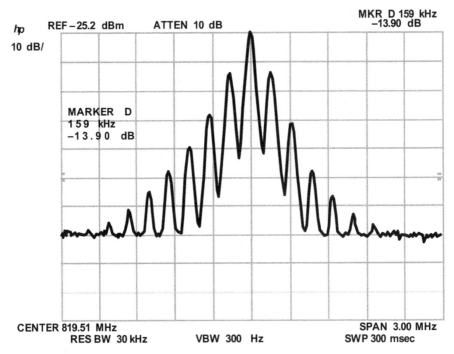

Figure 6.33 Measured output spectrum of a fractional-*N* frequency synthesizer with loop divisor $N = 160 + 1/32$ and the comparison frequency $f_r/R = 5.12$ MHz.

A fractional-*N* synthesizer achieves fine step size and low in-band phase noise with the penalty of fractional spurious tones, which comes from the periodic division ratio variation. Fractional spurs may be removable by using a high-order loop transfer function if the closest spur is outside of the PLL bandwidth. Note that the spacing of the closest spur to the carrier is determined by the synthesizer step size. For a synthesizer with fine step size smaller than the transfer function bandwidth, it is thus practically impossible to remove fractional spurs by using a loop LPF. Reducing the loop bandwidth to combat the fractional spurs means that you have to pay the penalty of longer lock time and increased out-of-band phase noise due to the VCO, as will be discussed in later sections. Even if the closest spur is outside of the loop filter bandwidth, removing those spurs normally requires a high-order loop filter with sharp roll-off, which increases the complexity and cost of the synthesizer. To remove the fractional spurious components for a synthesizer with fine step size, the best solution is to employ a $\Sigma\Delta$ noise shaper in the fractional accumulator [2]. Their function is to break up the repeated patterns of the loop divisor time sequence without affecting its average division ratio. This will result in the reduction or elimination of the spurs in spectrum.

6.9 All-Digital Phase Locked Loops

Traditionally PLLs have been implemented in much the way that has been discussed in this chapter so far; however, as technologies progress towards deep submicron CMOS, it becomes harder to implement analog blocks like charge pumps.

Additionally loop filters usually require off chip capacitors and resistors which is not conducive to a fully integrated solution. Therefore recently new PLL architectures are becoming popular which seek to replace much of the analog circuitry with digital equivalents [67–72]. This section will discuss in more detail these new ways of implementing frequency references.

6.9.1   The Evolution to a More Digital Architecture

We would like to take the more traditional PLL architecture shown in Figure 6.7 and make it more digital. A first step towards doing this would be to simply replace the analog loop filter and charge pump with a digital loop filter as shown in Figure 6.34. Here the PFD could also be thought of as a time-to-digital converter as it takes the phase difference and converts it to a digital signal (UP or DOWN). This signal is then lowpass-filtered in the digital domain by a discrete time filter shown here as $F(z)$. The digital output from this filter would then have to be fed through a DAC to create a voltage to drive the VCO circuit. The rest of the loop still remains the same at this stage.

The PLL can be made even more digital as shown in Figure 6.35. The VCO can be replaced by a digitally controlled oscillator removing the need for the DAC. This oscillator has bits at the input which determine which one of a number of discrete frequencies the oscillator will produce. Next the divider can be replaced by a time-to-digital converter (TDC). This block compares the timing of the DCO output edges to the crystal oscillator edges and produces a digital signal that is a measure of the DCO frequency. This is compared to the desired frequency set with the frequency control word (FCW), which is a digital input. The result is integrated to change frequency to phase and the result is passed to the digital loop filter to close the loop. This architecture then forms the basis of most all-digital phase locked loops.

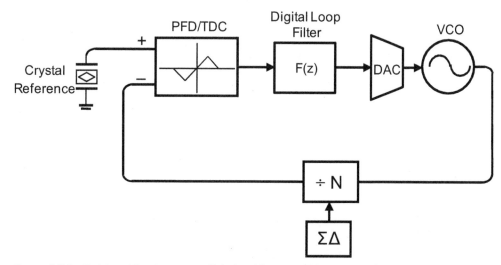

Figure 6.34 First transition to a more digital architecture. This removes the charge pump and replaces it and the loop filter with digital equivalents.

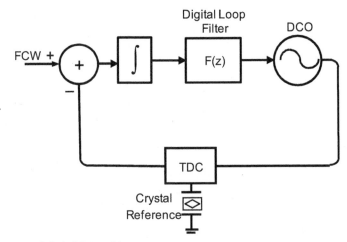

Figure 6.35 Basic all digital PLL architecture.

6.9.2 Phase Noise Limits Due to TDC Resolution

A time-to-digital converter is a digital circuit and therefore must be implemented with a finite number of bits and therefore a finite resolution. The TDC can only resolve the timing of the edges to T_{res}, the resolution of the TDC; thus, the rms timing error T_{error} is

$$T_{error}^2 = \frac{T_{res}^2}{12} \tag{6.84}$$

Note here that as with any linear converter the rms error is one-twelfth the step size. Therefore, the phase error will be:

$$\phi_n = 2\,\pi \frac{T_{error}}{T_{DCO}} \tag{6.85}$$

This noise will be spread from dc to F_{ref} (the sampling frequency) therefore the phase noise will be:

$$L = \left(2\pi \frac{T_{error}}{T_{DCO}}\right)^2 \frac{1}{F_{ref}} = \frac{(2\pi)^2}{12} \cdot \frac{T_{res}^2 F_{DCO}^2}{F_{ref}} \tag{6.86}$$

Therefore the in band phase noise plateau in (dBc/Hz) is

$$L(s) = 10\log\left[\frac{(2\pi)^2}{12} \cdot \frac{T_{res}^2 F_{DCO}^2}{F_{ref}}\right] \tag{6.87}$$

Example 6.8:   In-Band Phase Noise Limits in All-Digital Phase Lock Loops

An all-digital PLL is designed to have a DCO output frequency of 5 GHz, with a crystal reference frequency of 40 MHz. If the TDC used in the loop has a resolution of 20 ps, what is the best possible in-band phase noise that could be expected from this design?

Solution:

To answer this question, we need only make use of (6.87). Solving this equation with the numbers given in this example gives $L(s) = -90.8$dBc/Hz, which is not bad (but certainly not a record-breaking number at the time of this writing). Note that this is the upper limit on how good this can be though. This result does not take into account any other source of noise other than timing resolution of the TDC.

6.9.3   Phase Noise Limits Due to DCO Resolution

A DCO has a digital input and therefore must be implemented to produce a finite number of frequencies and therefore a finite frequency resolution. The DCO can only produce an arbitrary output frequency to an accuracy of f_{res}, the resolution of the DCO; thus, the rms frequency error f_{error} is

$$f_{error}^2 = \frac{f_{res}^2}{12} \tag{6.88}$$

Note here that as with any linear converter the rms error is one-twelfth the step size. This error will last for one reference period; therefore, the rms phase error will be:

$$\phi_n = f_{error} T_{ref} \tag{6.89}$$

This noise will be spread from dc to F_{ref} (the sampling frequency); therefore, the phase noise will be

$$L = \left(f_{error} T_{ref}\right)^2 \cdot \frac{1}{F_{ref}} = \frac{1}{12} \cdot \frac{f_{res}^2}{F_{ref}^3} \tag{6.90}$$

Therefore the phase noise plateau in (dBc/Hz) is

$$L(s) = 10 \log \left[\frac{1}{12} \cdot \frac{f_{res}^2}{F_{ref}^3} \right] \tag{6.91}$$

Note this is only the phase noise due to the quantization noise. Even if the DCO experiences no switching and the input is constant the circuit will still have a raw phase noise due to the circuit itself.

6.9.4   Time-to-Digital Converter Architecture

A basic time-to-digital converter determines how many VCO cycles are present in a complete period of the reference clock. A basic block diagram is shown in Figure 6.36. It consists of an integer and fractional counter, a differentiator and a summing block. The integer counter counts how many complete DCO cycles there are between rising edges of the reference signal. The fractional counter provides better resolution and counts the additional fraction of a cycle of the DCO between rising edges of the reference.

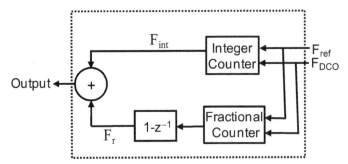

Figure 6.36 Functional block diagram of a basic time-to-digital converter.

An example of the operation of a TDC counting over one reference cycle is illustrated in Figure 6.37. After the rising edge of the reference the integer counter begins counting on the next rising edge of the DCO. It counts complete DCO cycles until the next rising edge of the reference and the count is labeled F_{int} in Figure 6.37.

If the reference and DCO edge do not happen to occur simultaneously then after the last complete DCO cycle and the rising edge of the reference edge there will be a fraction of a DCO cycle. The fractional counter counts the fraction of a period of the DCO between last DCO edge and the F_{ref} edge and is labeled F_{Rn} in Figure 6.37. In the previous period the same thing will have happened and that period the excess fraction of a DCO cycle right before the reference rising edge is labeled F_{Rn-1}. At the start of the reference cycle before the first rising edge of the DCO, the remainder of the DCO cycle is present labeled $1 - F_{Rn-1}$. So in one period, the excess period fraction F_r is

$$F_r = F_{Rn} + 1 - F_{Rn-1} = 1 + F_{Rn} - F_{Rn-1} = 1 + \left(1 - z^{-1}\right)F_{Rn} = \left(1 - z^{-1}\right)F_{Rn} \qquad (6.92)$$

Note that the "1" is a constant and can be omitted. Equation (6.92) explains the presence of the $1 - z^{-1}$ term in the block diagram in Figure 6.36.

Example 6.9:   Why Do We Need a Fractional Count?

An ADPLL is used with a FCW = 10.2 with only an integer counter. Note that provided there are sufficient bits at the input, any fractional multiple of the reference can be generated by an ADPLL; however, nothing comes for free. Determine the output of the TDC and its error after each reference cycle for six reference periods. Assume that the DCO is locked and running at a frequency of exactly 10.2 times the reference frequency.

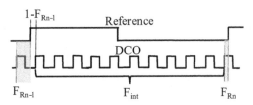

Figure 6.37 Illustrating integer and fractional count.

Solution:

Assuming that the reference and the DCO are in phase at the start of the experiment, then the waveforms for the first cycle are shown in Figure 6.38(a). After the first cycle of the reference, 10.2 cycles of the DCO have passed. An integer counter would have counted 10 cycles and therefore would have made an error of 0.2. The end of the second reference cycle can be seen in Figure 6.38(b). At the end of the second reference, 20.4 DCO cycles have passed. The integer counter would have counted another 10 DCO cycles and would now have made an error of 0.4. In this same way the results of six reference cycles can be analyzed and the results are summarized in Table 6.3. Note that as shown in Figure 6.39 the fifth cycle captures 11 rather than 10 DCO rising edges and therefore counts 11. After the fifth cycle the pattern repeats. As expected, four-fifths of the time the counter counts 10 and one-fifth of the time it counts 11. The average is 10.2.

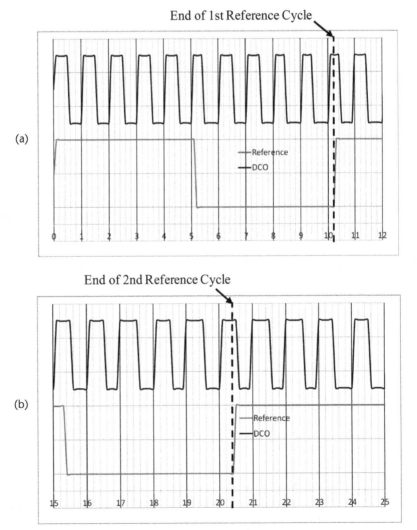

Figure 6.38   A DCO running 10.2 times faster than the reference: (a) the first cycle and (b) the completion of the second cycle.

Table 6.3   Results of Using an Integer Counter for Five Reference Cycles to Produce a Ratio of 10.2

Reference Cycle	Integer Count	Error	DCO Cycles Since Start
1	10	0.2	10.2
2	10	0.4	20.4
3	10	0.6	30.6
4	10	0.8	40.8
5	11	0.0	51
6	10	0.2	61.2

Notice in the previous example that the value of the error repeats every five reference cycles. This is due to the fact that the fraction in this case is 1/5. In general, we can expect that strong spurs would be present at a frequency of

$$f_{\text{spur}} = F_r \cdot F_{\text{ref}} \tag{6.93}$$

Now if a fractional counter is used in theory this could make this error go away, provided that infinite resolution is used. With finite resolution there will always be some error although it can be much smaller than with an integer counter.

6.9.5   The Digital Loop Filter

Previously in an analog loop the basic loop filter consisted of a resistor and capacitor as shown in Figure 6.40(a). In this case the transfer function for the filter was found to be

$$F(s) = R + \frac{1}{sC_1} \tag{6.94}$$

Looking at the transfer function, it should be straightforward to draw a block diagram for this transfer function and one is shown in Figure 6.40(b). A generic s

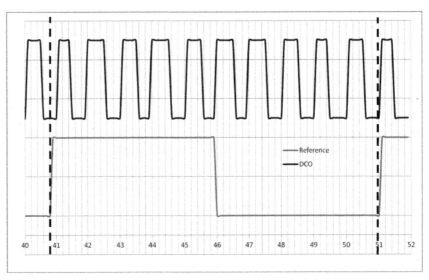

Figure 6.39   The fifth cycle of a DCO running 10.2 times faster than the reference. In this cycle 11 rather than 10 rising DCO references are caught between the two reference edges.

domain function with the same form as Figure 6.40(b) is shown in Figure 6.40(c). Here the integral path is scaled by T_{Ref} and the transfer function for this block diagram is

$$F(s) = K_{prop} + \frac{K_{int}}{sT_{ref}} \qquad (6.95)$$

Replacing the s-domain functions with the equivalent z-domain functions gives a discrete time filter function as shown in Figure 6.40(d). T_{ref} was included to denormalize the loop filter. The transfer function of the digital equivalent is

$$F(z) = K_{prop} + K_{int}\left(\frac{1}{z-1}\right) \qquad (6.96)$$

Note that if you had already designed an analog loop filter in this case for the same frequency response:

$$K_{prop} = R$$

$$K_{int} = \frac{T_{ref}}{C_1} \qquad (6.97)$$

This digital loop filter may be placed into the loop as shown in Figure 6.41. Here the loop gain may also be adjusted by use of a gain block in the feedforward path called LG. The question of how the digital circuits are clocked still needs to be answered as there is no clock with analog circuits. The answer is that the digital circuits must be synchronized with the reference signal and the reference can be used to provide the reference for the digital logic.

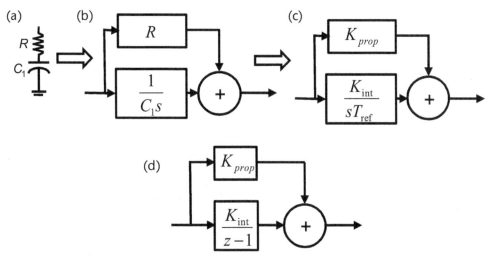

Figure 6.40 (a) An analog loop filter, (b) a block diagram of equivalent of (a), (c) a more generic version of (b), and (d) a digital equivalent of (c).

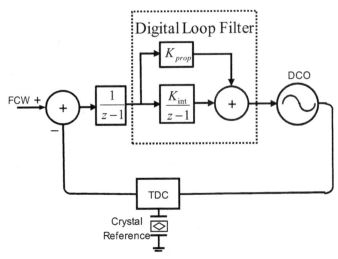

Figure 6.41   ADPLL with a basic digital loop filter.

Example 6.10:   A Simple ADPLL Design

Take the PLL of Example 6.2 and convert it into an ADPLL design with a simple integer TDC. Show simulation results for a verity of control words. For this example, assume that the DCO and digital circuitry has infinite resolution.

Solution:
In that example an output frequency of 4 GHz was to be synthesized using a reference frequency of 40 MHz. C_1 was found to be 5.66 nF, R was 530Ω, K_{VCO} was 200 MHz/V, and K_{phase} was 100 μA/rad. We now implement the loop as shown in Figure 6.42 with F_{Ref} = 40 MHz, K_{prop} = R = 530, and K_{int} = T_{ref}/C_1 = 4.42. Note that to account for the gain of the divider in the analog loop, a gain block of value 1/FCW is used, and the output of the integrator is multiplied by 2π to convert the result to radians. Then the same value for K_{phase} can be used as in the previous example. Also note that the sensitivity of the DCO in this example will be 200 MHz/unit (a unit less number is now fed into the DCO instead of a voltage) to make it the same as the previous example.

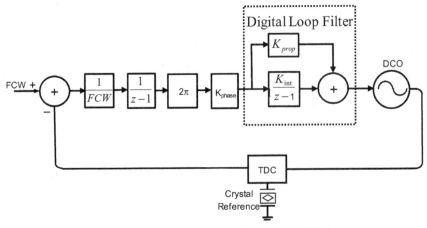

Figure 6.42   ADPLL with its loop gain adjusted to match a fractional-*N* design.

First, to verify that the loop operates correctly, a simulation is run where the input FCW is stepped from 100 to 102 as shown in Figure 6.43. In this figure the control signal value going into the DCO is plotted. Note that with a gain of 200 MHz/unit the control signal must move from zero to 0.4 to move the output frequency by 80 MHz to 4.08 GHz. Note the similarity in the settling characteristic to Example 6.2. As the loop parameters are the same, we would expect the settling characteristics of both designs to be very similar.

It is also interesting to look at the error signal when the loop is set to a non-integer FCW value. Here we plot the error signal for FCW = 100.5 in Figure 6.44 and FCW = 100.2 in Figure 6.45. In each simulation the center frequency of the DCO is set so that an input of zero will center the DCO at the desired output frequency. When FCW =100.5, the error signal forms a square wave with a frequency of 20 MHz (half the reference frequency). Note that the waveform is not a perfect square wave and sometimes "skips" a bit. This is due to other imperfections in the feedback loop which periodically need adjusting. When FCW = 100.2, the output is a square wave with an 80% duty cycle and a frequency of 8 MHz (one-fifth the reference frequency). The average value is still zero to give the desired output frequency.

While it is known that spurious performance of a ADPLL with only an integer TDC will not be good, an FFT of the DCO output when FCW = 100.5 is shown in Figure 6.46. Here the desired output frequency of 4.02 GHz is shown at the center of the plot. There are two distinct tones at 4.0 and 4.4 GHz (each 20 MHz away from the center frequency). A second set of tones at 4.08 and 3.96 GHz (60 MHz away from the center frequency) can also be seen. These tones would correspond to the third harmonic of the control signal.

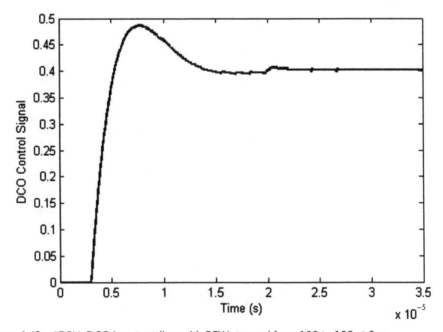

Figure 6.43   ADPLL DCO input settling with FCW stepped from 100 to 102 at 3 μs.

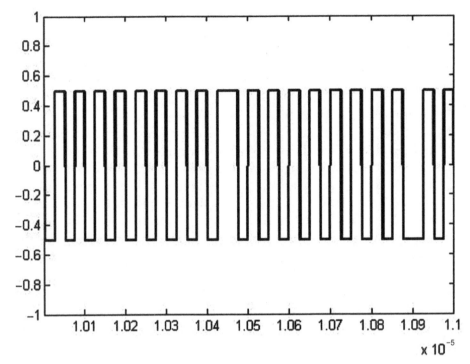

Figure 6.44   ADPLL error signal with FCW = 100.5.

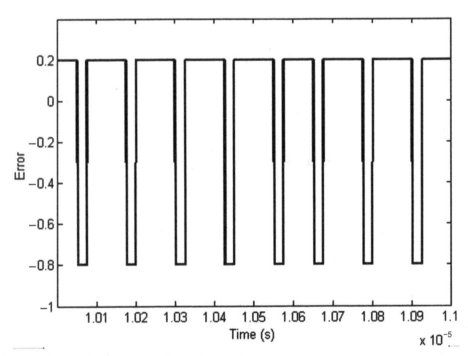

Figure 6.45   ADPLL error signal with FCW = 100.2.

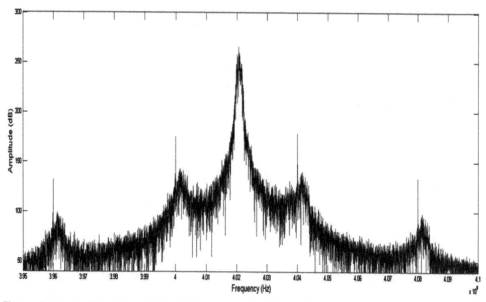

Figure 6.46   An FFT of the ADPLL DCO output with FCW = 100.5.

6.9.6   ADPLL Noise Calculations

An ADPLL model for doing noise calculations is shown in Figure 6.47. Because a digital or clocked system looks almost identical to a continuous one provided that $f < 1/10\ f_{\text{ref}}$, all the previous theory developed in Section 6.6.4 can be reused here to deal with raw phase noise from the DCO as well as noise from the crystal oscillator. Note that the TDC and the integrator have been replaced with a standard summing block and that the s domain equivalent of the loop filter may be placed into the calculation for simplicity. Note that in this case K_{DCO} must be determined. For a linear DCO one can assume that K_{DCO} will be equal to the frequency step size and will have units of frequency because in this case the denominator is a simple number rather than a voltage.

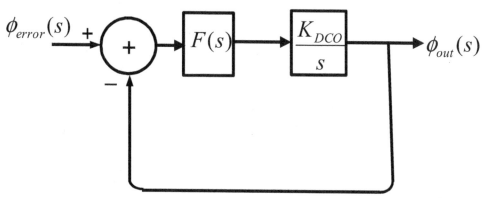

Figure 6.47   ADPLL model for noise calculations.

6.9.7 Time-to-Digital Converter Circuits

An example of a basic TDC circuit is shown in Figure 6.48. Here the DCO signal is connected to a string of inverter circuits each with delay τ. The inverters are connected to the D input of D flip flops and the clock input is connected to the reference signal. It is easiest to see how this circuit operates by first considering the case where the DCO is ahead of the reference. On the rising edge of the DCO the input to the first inverter goes high and after a delay of τ its output goes high as well. Thus, if the reference edge is late by $N\tau$, then the first N inverters will be high and the rest will be low. Once the reference edge arrives the value of the inverters in the chain is stored in the D flip flops. By counting how many flip flops have a high output we can determine by how many inverter delays the DCO was early by. If enough inverters are used, such that at least one complete DCO cycle is captured by the TDC, it will always be possible to determine how early or late the DCO edge was.

Example 6.11:   Designing the TDC and Determining the TDC Circuit Output

An inverter is available with a $\tau \approx 1/10\ T_{DCO}$. Design and determine the output of a TDC when the DCO is early by τ and 2τ, when the DCO is late by τ and 2τ, and when the DCO is in phase with the reference.

Solution:
Because $\tau = 1/10\ T_{DCO}$, 10 delay elements are needed to cover a complete cycle. Assuming the DCO and reference are in phase, the output of the flip flops will be 0000011111. The five 1s will be due to the previous half-cycle when the DCO was high, which is still making its way through the last five inverters. For the case where the DCO is early by τ, the output will be 1000001111. In this case only the first inverter will go high before the reference. For the case where the DCO is early by 2τ the output will be 1100000111. When the DCO is late by τ, the output will be 0000111110; in this case, we can see that the previous cycle high has not made it far enough through the inverters to turn the last inverter high. When the DCO is late by 2τ, the last two inverters will be low, giving 0001111100.

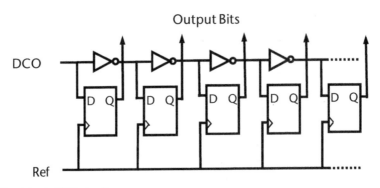

Figure 6.48   A basic TDC circuit.

Note that the pattern of bits can also be thought of as a picture of the DCO waveform sampled once every τ and measured relative to the reference.

By examining the bits that come out of a TDC the number of delays early or late the DCO is can be determined. The length of a DCO period in delays can also be measured. Knowing the length of the period in delays t_p and the number of delays early or late t_r that the DCO is then the faction of a period can be computed as:

$$F_{Rn} = \frac{t_r}{t_p} \tag{6.98}$$

An alternative TDC circuit called a Vernier TDC, which uses two delay lines with unequal delays, is shown in Figure 6.49. In doing this, the resolution can be improved to the limit of the difference in the two delays rather than the absolute value of a single delay. In this circuit two delay lines are used to feed a set of D flip flops. The difference between the two delays is given by

$$\Delta = \tau_1 - \tau_2 \tag{6.99}$$

and is the resolution of the Vernier TDC.

When a signal is fed into a Vernier TDC, as shown in Figure 6.49, a basic Vernier TDC, the distance between the rising edges of the two signals will decrease by Δ as they pass through each delay stage. At some point along the chain, the DCO edge will fall behind the reference edge and at that point the output of the flip flops will change from high to low. The number of stages it takes for this to happen tells you how many Δs the DCO is ahead of the reference. This circuit has the potential for higher resolution at the cost of an additional delay line.

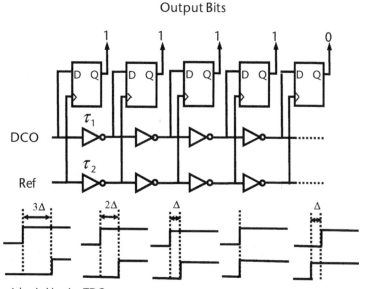

Figure 6.49   A basic Vernier TDC.

References

[1]   Larson, L. E., (ed.), *RF and Microwave Circuit Design for Wireless Communications*, Norwood, MA: Artech House, 1997.

[2]   Rogers, J., C. Plett, and F. Dai, *Integrated Circuit Design for High-Speed Frequency Synthesis*, Norwood, MA: Artech House, 2006.

[3]   Craninckx, J. and M. Steyaert, *Wireless CMOS Frequency Synthesizer Design*, Dordrecht, The Netherlands: Kluwer Academic Publishers, 1998.

[4]   Kroupa, V. F., "Low-Noise Microwave Synthesizer Design Principles, in *Direct Digital Frequency Synthesizers*, New York: IEEE Press, 1999, pp. 175–180.

[5]   Manassewitch, V., *Frequency Synthesizers Theory and Design,* New York: John Wiley & Sons, 2005.

[6]   Noordanus, J., "Frequency Synthesizers—A Survey of Techniques," *IEEE Transactions on Communications Techniques*, Vol. 17, No. 2, April 1969, pp. 257–271.

[7]   Reinhardt, V. S. et al., "A Short Survey of Frequency Synthesizer Techniques," *Frequency Control Symposium,* May 1986, pp. 355–365.

[8]   Rohde, U. L., *Digital Frequency Synthesizers: Theory and Design*, Upper Saddle River, NJ: Prentice Hall, 1983.

[9]   Rohde, U. L., *Microwave and Wireless Synthesizers: Theory and Design*, New York: John Wiley & Sons, 1997.

[10]   Lindsay, W. C., *Phase-Locked Loops*, New York: IEEE Press, 1986.

[11]   Razavi, B., *Design of Integrated Circuits for Optical Communications*, New York: McGraw-Hill, 2002.

[12]   Razavi, B., *Monolithic Phase-Locked Loops and Clock Recovery Circuits*, New York: Wiley-IEEE Press, 1996.

[13]   Best, R. E., *Phase-Locked Loops: Theory, Design, and Applications*, 6th ed., Boston, MA: McGraw-Hill, 2007.

[14]   Blanchard, A., *Phase-Locked Loops: Applications to Coherent Receiver Design*, New York: John Wiley & Sons, 1976.

[15]   Gardner, F. M., *Phaselock Techniques*, New York: John Wiley & Sons, 1979.

[16]   Egan, W. F., *Frequency Synthesis by Phase Lock*, New York: John Wiley & Sons, 2000.

[17]   Kroupa, V. F., "Noise Properties of PLL Systems," *IEEE Transactions on Communications*, Vol. 30, October 1982, pp. 2244–2252.

[18]   Wolaver, D. H., *Phase-Locked Loop Circuit Design*, Upper Saddle River, NJ: Prentice Hall, 1991.

[19]   Crawford, J. A., *Frequency Synthesizer Design Handbook*, Norwood, MA: Artech House, 1994.

[20]   Lee, H. et al., "A Σ-Δ Fractional-N Frequency Synthesizer Using a Wide-Band Integrated VCO and a Fast AFC Technique for GSM/GPRS/WCDMA Applications," *IEEE J. Solid-State Circuits*, Vol. 39, July 2004, pp. 1164–1169.

[21]   Leung, G., and H. Luong, "A 1-V 5.2 GHz CMOS Synthesizer for WLAN Applications," *IEEE J. Solid-State Circuits*, Vol. 39, November 2004, pp. 1873–1882.

[22]   Rhee, W., B. Song, and A. Ali, "A 1.1- GHz CMOS Fractional-N Frequency Synthesizer with a 3-b Third-Order ΣΔ Modulator," *IEEE J. Solid-State Circuits*, Vol. 35, October 2000, pp. 1453–1460.

[23]   Lo, C., and H. Luong, "A 1.5-V 900-MHz Monolithic CMOS Fast-Switching Frequency Synthesizer for Wireless Applications," *IEEE J. Solid-State Circuits*, Vol. 37, April 2002, pp. 459–470.

[24]   Ahola, R., and K. Halonen, "A 1.76- GHz 22.6mW ΔΣ Fractional-N Frequency Synthesizer," *IEEE J. Solid-State Circuits*, Vol. 38, January 2003, pp. 138–140.

[25] Heng, C., and B. Song, "A 1.8- GHz CMOS Fractional-N Frequency Synthesizer with Randomizer Multiphase VCO," *IEEE J. Solid-State Circuits*, Vol. 38, June 2003, pp. 848–854.

[26] Park, C., O. Kim, and B. Kim, "A 1.8-GHz Self-Calibrated Phase-Locked Loop with Precise I/Q Matching," *IEEE J. Solid-State Circuits*, Vol. 36, May 2001, pp. 777–783.

[27] Klepser, B., M. Scholz, and E. Götz, "A 10-GHz SiGe BiCMOS Phase-Locked-Loop Frequency Synthesizer," *IEEE J. Solid-State Circuits*, Vol. 37, March 2002, pp. 328–335.

[28] Pellerano, S. et al., "A 13.5-mW 5-GHz Frequency Synthesizer with Dynamic-Logic Frequency Divider," *IEEE J. Solid-State Circuits*, Vol. 39, February 2004, pp. 378–383.

[29] Leenaerts, D. et al., "A 15-mW Fully Integrated I/Q Synthesizer for Bluetooth in 0.18-μm CMOS," *IEEE J. Solid-State Circuits*, Vol. 38, July 2003, pp. 1155–1162.

[30] Kan, T., G. Leung, and H. Luong, "A 2-V 1.8-GHz Fully Integrated CMOS Dual-Loop Frequency Synthesizer," *IEEE J. Solid-State Circuits*, Vol. 37, August 2002, pp. 1012–1020.

[31] Aytur, T., and B. Razavi, "A 2-GHz, 6-mW BiCMOS Frequency Synthesizer," *IEEE J. Solid-State Circuits*, Vol. 30, December 1995, pp. 1457–1462.

[32] Chen, W. et al., "A 2-V 2.3/4.6-GHz Dual-Band Frequency Synthesizer in 0.35-μm Digital CMOS Process," *IEEE J. Solid-State Circuits*, Vol. 39, January 2004, pp. 234–237.

[33] Yan, W., and H. Luong, "A 2-V 900-MHz CMOS Dual-Loop Frequency Synthesizer for GSM Receivers," *IEEE J. Solid-State Circuits*, Vol. 36, February 2001, pp. 204–216.

[34] Shu, K. et al., "A 2.4-GHz Monolithic Fractional-N Frequency Synthesizer with Robust Phase-Switching Prescaler and Loop Capacitance Multiplier," *IEEE J. Solid-State Circuits*, Vol. 38, June 2003, pp. 866–874.

[35] Shu, Z., K. Lee, and B. Leung, "A 2.4-GHz Ring-Oscillator-Based CMOS Frequency Synthesizer with a Fractional Divider Dual-PLL Architecture," *IEEE J. Solid-State Circuits*, Vol. 39, March 2004, pp. 452–462.

[36] McMahill, D., and C. Sodini, "A 2.5-Mb/s GFSK 5.0-Mb/s 4-FSK Automatically Calibrated Σ-Δ Frequency Synthesizer," *IEEE J. Solid-State Circuits*, Vol. 37, January 2002, pp. 18–26.

[37] Lam, C., and B. Razavi, "A 2.6-GHz/5.2-GHz Frequency Synthesizer in 0.4-μm CMOS Technology," *IEEE J. Solid-State Circuits*, Vol. 35, May 2000, pp. 788–794.

[38] Perrott, M., T. Tewksbury, and C. Sodini, "A 27-mW CMOS Fractional-N Synthesizer Using Digital Compensation for 2.5-Mb/s GFSK Modulation," *IEEE J. Solid-State Circuits*, Vol. 32, December 1997, pp. 2048–2060.

[39] Temporiti, E. et al., "A 700-kHz Bandwidth $\Sigma\Delta$ Fractional Synthesizer with Spurs Compensation and Linearization Techniques for WCDMA Applications," *IEEE J. Solid-State Circuits*, Vol. 39, September 2004, pp. 1446–1454.

[40] Dehng, G. et al., "A 900-MHz 1-V CMOS Frequency Synthesizer," *IEEE J. Solid-State Circuits*, Vol. 35, August 2000, pp. 1211–1214.

[41] Lin, T., and W. Kaiser, "A 900-MHz 2.5-mA CMOS Frequency Synthesizer with an Automatic SC Tuning Loop," *IEEE J. Solid-State Circuits*, Vol. 36, March 2001, pp. 424–431.

[42] Rategh, H., H. Samavati, and T. H. Lee, "A CMOS Frequency Synthesizer with an Injection-Locked Frequency Divider for a 5-GHz Wireless LAN Receiver," *IEEE J. Solid-State Circuits*, Vol. 35, May 2000, pp. 780–787.

[43] Hwang, I., S. Song, and S. Kim, "A Digitally Controlled Phase-Locked Loop with a Digital Phase-Frequency Detector for Fast Acquisition," *IEEE J. Solid-State Circuits*, Vol. 36, October 2001, pp. 1574–1581.

[44] Zhang, B., P. Allen, and J. Huard, "A Fast Switching PLL Frequency Synthesizer with an On-Chip Passive Discrete-Time Loop Filter in 0.25-μm CMOS," *IEEE J. Solid-State Circuits*, Vol. 38, October 2003, pp. 855–865.

[45] Da Dalt, N. et al., "A Fully Integrated 2.4-GHz LC-VCO Frequency Synthesizer with 3-ps Jitter in 0.18-μm Standard Digital CMOS Copper Technology," *IEEE J. Solid-State Circuits*, Vol. 37, July 2002, pp. 959–962.

[46]   Craninckx, J., and M. S. J. Steyaert, "A Fully Integrated CMOS DCS-1800 Frequency Synthesizer," *IEEE J. Solid-State Circuits*, Vol. 33, December 1998, pp. 2054–2065.

[47]   Koo, Y. et al., "A Fully Integrated CMOS Frequency Synthesizer with Charge-Averaging Charge Pump and Dual-Path Loop Filter for PCS- and Cellular-CDMA Wireless Systems," *IEEE J. Solid-State Circuits*, Vol. 37, May 2002, pp. 536–542.

[48]   Bax, W. T., and M. A. Copeland, "A GMSK Modulator Using a $\Delta\Sigma$ Frequency Discriminator-Based Synthesizer," *IEEE J. Solid-State Circuits*, Vol. 36, August 2001, pp. 1218–1227.

[49]   Bietti, I. et al., "An UMTS $\Sigma\Delta$ Fractional Synthesizer with 200kHz Bandwidth and -128dBc/Hz @ 1 MHz Using Spurs Compensation and Linearization Techniques," *Proc. IEEE Custom Integrated Circuits Conference*, 2003, pp. 463–466.

[50]   De Muer, B., and M. Steyaert, "A CMOS Monolithic $\Delta\Sigma$-Controlled Fractional-N Frequency Synthesizer for DCS-1800," *IEEE J. Solid-State Circuits*, Vol. 37, July 2002, pp. 835–844.

[51]   Riley, T. A., M. Copeland, and T. Kwasniewski, "Delta–Sigma Modulation in Fractional-N Frequency Synthesis," *IEEE J. Solid-State Circuits*, Vol. 28, May 1993, pp. 553–559.

[52]   Rogers, J. W. M. et al., "A $\Delta\Sigma$ Fractional-N Frequency Synthesizer with Multi-Band PMOS VCOs for 2.4 and 5 GHz WLAN Applications," *European Solid-State Circuits Conference (ESSCIRC)*, September 2003, pp. 651–654.

[53]   Vaucher, C. S. et al., "A Family of Low-Power Truly Modular Programmable Dividers in Standard 0.35μm CMOS Technology," *IEEE J. Solid-State Circuits*, Vol. 35, July 2000, pp. 1039–1045.

[54]   Razavi, B., *RF Microelectronics*, Upper Saddle River, NJ: Prentice Hall, 1998.

[55]   Kroupa, V. F., "Jitter and Phase Noise in Frequency Dividers," *IEEE Transactions on Instrumentation and Measurement*, Vol. 50, No. 5, October 2001, p. 1241.

[56]   Papoulis, A., *Probability, Random Variables, and Stochastic Processes,* New York: McGraw-Hill, 1984.

[57]   Sze, S. M., *Physics of Semiconductor Devices,* 2nd ed., New York: John Wiley & Sons, 1981.

[58]   Gray, P. R. et al., *Analysis and Design of Analog Integrated Circuits,* 4th ed., New York: John Wiley & Sons, 2001.

[59]   Amaya, R. E. et al., "EM and Substrate Coupling in Silicon RFICs," *J. Solid State Circuits*, Vol. 40, No. 9, September 2005.

[60]   Leeson, D. B., "A Simple Model of Feedback Oscillator Noise Spectrum," *Proc. IEEE*, February 1966, pp. 329–330.

[61]   Uwano, T. et al., "Design of a Low-Phase Noise VCO for an Analog Cellular Portable Radio Application," *Electronics and Communications in Japan*, Vol. 77, 1994, pp. 58–65.

[62]   Dow, S. et al., "A Dual-Band Direct-Conversion Transceiver IC for GSM," *Proc. International Solid-State Circuits Conference*, 2002, pp. 230–462.

[63]   Vig, J. R., "Quartz Crystal Resonators and Oscillators," U.S. Army Electronics Technology and Devices Report, SLCET-TR-88-1, 1988.

[64]   Watanabe, Y. et al., "Phase Noise Measurements in Dual-Mode SC-Cut Crystal Oscillators," *IEEE Transactions on Ultrasonics, Ferroelectrics and Frequency Control*, Vol. 47, No. 2, March 2000, pp. 374–378.

[65]   Rohde, U. L., "A Novel RFIC for UHF Oscillators," *IEEE Radio Frequency Integrated Circuits (RFIC) Symposium*, 2000, pp. 53–56.

[66]   Kivinen, J., and P. Vainikainen, "Phase Noise in a Direct Sequence Based Channel Sounder," *8th IEEE International Symposium on Personal, Indoor and Mobile Radio Communications*, Vol. 3, 1997, pp. 1115–1119.

[67]   Staszewski, R. B. et al., "All-Digital PLL and Transmitter for Mobile Phones," *IEEE J. Solid-State Circuits*, Vol. 40, December 2005, pp. 2469–2482.

[68] Syllaios, I. L., R. B. Staszewski, and P. T. Balsara, "Time-Domain Modeling of an RF All-Digital PLL," *IEEE Transactions on Circuits and Systems II*, Vol. 55, June 2008, pp. 601–605.

[69] Staszewski, R. B. et al., "TDC-Based Frequency Synthesizer for Wireless Applications," *IEEE Radio Frequency Integrated Circuits Symposium*, 2004, pp. 215–218.

[70] Staszewski, R. B., and P. T. Balsara, *All-Digital Frequency Synthesizer in Deep-Submicron CMOS*, New York: John Wiley & Sons, 2006.

[71] Liscidini, A., L. Vercesi, and R. Castello, "Time to Digital Converter Based on a 2-Dimensions Vernier Architecture," *IEEE Custom Integrated Circuits Conference*, 2009, pp. 45–48.

[72] Temporiti, E. et al., "Insights into Wideband All-Digital PLLs for RF Applications," *IEEE Custom Integrated Circuits Conference*, 2009, pp. 37–44.

CHAPTER 7

Block-Level Radio Design Examples

7.1 An IEEE 802.11n Transceiver for the 5-GHz Band

Wireless local area networks (WLANs) are now ubiquitous, and nearly every computer, laptop, tablet, and cell phone comes equipped with a WLAN interface. Most of these devices implement the IEEE 802.11 standard, which, among many other things, describes the air interface for WLAN equipment [1]. The most recent amendment to the IEEE 802.11 standard, IEEE 802.11n, provides several improvements [2]. In particular, it supports the use of multiple transmit and receive antennas to deliver higher throughput and greater range. The air interface is based on OFDM transmission with either 20-MHz or 40-MHz channels in the 2.4-GHz and 5-GHz unlicensed bands. To describe a radio that meets all the requirements for WLANs would be hard to cover as only part of one chapter in a book, but here we will at least try to sketch out some of the requirements for a radio to be used with a 20-MHz channel in the 5-GHz band. Before describing the RF components in detail, we present a brief introduction to the required baseband signal processing.

7.1.1 Baseband Signal Processing

The IEEE 802.11n amendment with a 20-MHz channel calls for the use of OFDM transmission with 56 subcarriers, although only 52 are used to carry data and the other four contain pilot symbols used for carrier synchronization. Four different modulation schemes (BPSK, 4-QAM, 16-QAM, and 64-QAM) are supported, along with four different error correcting code rates (1/2, 2/3, 3/4, and 5/6). Eight different combinations of the modulation schemes and code rates are defined in the standard, as outlined in Table 7.1. The OFDM symbol duration is 4 µs, and the number of message bits that are transmitted with each symbol is shown in Table 7.1, along with the data transmission rate. The packet error rate for a packet size of 4,096 bytes, as a function of the SNR, for each of these schemes is shown in Figure 7.1. The system dynamically changes which modulation and coding scheme (MCS) is used in response to changing channel conditions to try to get the highest throughput possible while providing acceptably low error probabilities.

The amendment allows for the use of up to four antennas at the transmitter and receiver. If the transmitter and receiver have the same number of antennas, then the system can use spatial multiplexing to simultaneously transmit multiple streams of data, where the number of streams is equal to the number of antennas. This allows for up to four times greater throughput than what is specified in Table 7.1. If the receiver has more antennas than the transmitter, the receiver can use the

Table 7.1   IEEE 802.11n Modulation and Coding Schemes (MCS)

MCS Index	Modulation	Code Rate	Number of Message Bits per OFDM Symbol	Data Rate (Mbps)
0	BPSK	1/2	26	6.5
1	4-QAM	1/2	52	13.0
2	4-QAM	3/4	78	19.5
3	16-QAM	1/2	104	26.0
4	16-QAM	3/4	208	39.0
5	64-QAM	2/3	208	52.0
6	64-QAM	3/4	234	58.5
7	64-QAM	5/6	260	65.0

extra antennas for receive diversity to provide better protection against fading and to increase the effective range of the system. However, if there are more transmit antennas than receive antennas, the transmitter can use space-time block codes or beamforming to provide transmit diversity.

Transmission is carried out on variable length data packets containing up to 65,535 bytes. Generation of the transmitted baseband signal involves several steps, including scrambling, encoding, spatial parsing, interleaving, symbol mapping, space-time block coding, spatial mapping, inverse discrete Fourier transforming, cyclic prefix insertion, and digital-to-analog conversion, as shown in Figure 7.2. These steps are briefly described next.

The data is scrambled by finding the bit-by-bit exclusive-OR of the data with a deterministic pseudo-random binary scrambling sequence. Descrambling at the receiver is possible by repeating the process with a locally generated version of the scrambling sequence. Scrambling reduces the likelihood of long sequences of zeros and ones in the output, which can have detrimental effects on the PSD of the transmitted signal and make carrier recovery at the receiver less reliable.

The scrambled data field is then encoded with an error correcting code. The redundancy introduced at the output of the encoder can be exploited by the receiver

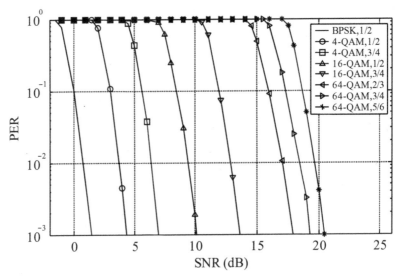

Figure 7.1   Packet error rate versus SNR for IEEE 802.11n modulation and coding schemes.

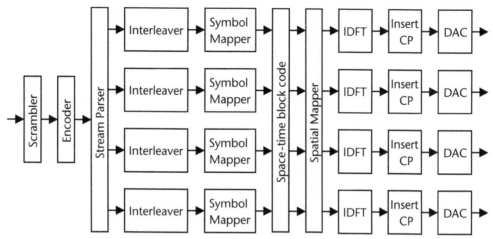

Figure 7.2   Block diagram of the baseband signal processing components of the transmitter.

to correct for some transmission errors, yielding more reliable communication. Either a binary convolutional code or a low density parity check (LDPC) code can be used, and code rates of 1/2, 2/3, 3/4, and 5/6 are supported. Lower-rate codes provide better protection against errors, at the expense of additional transmission overhead.

The code bits may then be split into different spatial streams by the stream parser, provided that the transmitter and receiver have enough antennas. The spatial streams are transmitted simultaneously, thereby increasing the system throughput. The number of spatial streams cannot exceed the number of antennas at the transmitter or the receiver, whichever is smaller.

When convolutional coding is used, the code bits for each spatial stream are then rearranged by an interleaver. Because adjacent code bits at the output of the convolutional encoder are highly correlated, interleaving is necessary to reduce the correlation between bits transmitted in the same signal constellation symbol and on the adjacent subcarriers. Note that interleaving is not necessary when LDPC codes are used, because adjacent code bits are not very correlated.

The interleaved bits are then mapped to points in the signal constellation. The following signal constellations are available for use: BPSK, 4-QAM, 16-QAM, and 64-QAM. The higher-order modulation schemes provide better throughput, but require a higher signal-to-noise ratio at the receiver. All the symbols transmitted in any one spatial stream use the same modulation scheme, but different spatial streams can use different modulation schemes.

If the number of spatial streams is less than the number of transmit antennas, then it is possible to use space-time block coding to provide for more robust transmission. In particular, the Alamouti STBC described in Chapter 2 can be used to provide transmit diversity. The symbols in the space-time streams are rearranged (and some symbols are duplicated), to provide up to four space-time streams, one for each transmit antenna.

Spatial mapping can then be performed on a per-subcarrier basis and involves a linear transformation of the symbols in the *space-time streams*. If accurate knowledge of the channel has been obtained by the transmitter (usually after channel sounding packets have been sent and the results reported back from the receiver), it is possible to use the spatial mapping to perform beamforming to better direct the transmitted energy towards the receiver. Spatial mapping can also be used for *spatial expansion*, as an alternative to space-time block coding, to make use of all the transmit antennas when it is not possible to use more spatial streams.

The modulation symbols produced by the symbol mapper, modified by the space-time block coding and spatial mapping for each transmit antenna are arranged into groups of 52 symbols, and symbols within a group are mapped to 52 of the 56 subcarriers. Deterministic pilot symbols are mapped to the other four subcarriers. If X_k is the modulation symbol transmitted on subcarrier k, the complex baseband transmitted signal for one OFDM symbol is ideally

$$v(t) = h_T(t) \sum_{\substack{k=-26 \\ k \neq 0}}^{26} X_k e^{j2\pi\Delta_F k(t-T_{GI})} \tag{7.1}$$

where

$$h_T(t) = \begin{cases} 1, & \text{if } 0 \leq t < T_{SYM} \\ 0, & \text{otherwise} \end{cases} \tag{7.2}$$

is the windowing function with a duration equal to the OFDM symbol interval of $T_{SYM} = 4$ μsec, $\Delta_F = 20/64 = 0.3125$ MHz is the subcarrier separation, and $T_{GI} = 0.8$ is the duration of the guard interval. A windowing function that is slightly longer than the duration of the OFDM symbol, and tapered at the ends, may be used to provide better spectral shaping, in which case adjacent OFDM symbols will overlap slightly.

In practice, the baseband signal is not generated directly according to (7.1), as this would require 52 different frequency synthesizers, which could be quite complicated. Instead, the inverse discrete Fourier transform (IDFT) of the modulation symbols is first calculated. The 56 modulation symbols (including pilot symbols) are fed into a 64-point IDFT at the input positions indicated in Figure 7.3, and zeros are applied to the other inputs. The output, which represents samples of $v(t)$ from (7.1), is then converted to a serial form. A replica of the last 16 samples of the OFDM symbol is placed before the samples, creating the cyclic prefix, which is

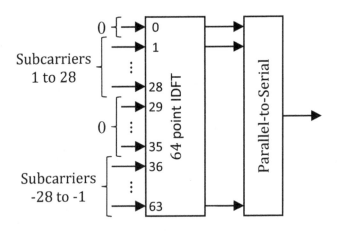

Figure 7.3   IDFT used to generate the time-domain OFDM symbol.

transmitted during the guard interval. Inclusion of the cyclic prefix allows for easier compensation for the delay spread of the channel. The samples with the cyclic prefix are then passed to a digital-to-analog converter to generate the transmitted baseband signal, which should be filtered with a lowpass filter and upconverted to the desired frequency band.

7.1.2   RF Considerations

In this section we will talk about some of the radio front-end considerations for a WLAN radio operating in the 5-GHz band only. While the 5-GHz band can vary depending on the country in which the radio is deployed, a fairly common range is 5.25 to 5.35 GHz. In this frequency band, the standard allows for a maximum transmit power of 200 mW (23 dBm), provided that the antenna gain is 6 dBi or less.

The documentation for WLAN provides instructions for receiver sensitivity operating with the various modulation schemes, and describes how much adjacent channel and alternate adjacent channel rejection are required. The transmitter EVM is also specified. These values, which depend on the modulation and coding scheme, are summarized in Table 7.2. The maximum receive power of signals in undesired channels is also stipulated as −30 dBm. Also an extremely important piece of information when designing a transmitter is to have the spectral mask. The spectral mask tells the designers how much power can be leaked into adjacent channels. The spectral mask is drawn in Figure 7.4.

The standard that we are considering is 802.11n. The n part of this standard allows for multiple antennas. Therefore, we will assume that there will be no deep fades and that an additional 25 dB of link margin isn't required. Also, we will assume that the entire 6 dBi of antenna gain can be realized by the link. Now let us do some simple link budget calculations. We will start by using a reference distance of 4m and a transmit and receive antenna gain of 3 dBi each for a total of 6 dBi. In this case, the reference received power would be (assuming line of sight and a carrier frequency of 5.25 GHz)

$$P_R = P_T \frac{G_T G_R}{(4\pi d/\lambda_c)^2} = 200 \text{ mW} \frac{(2)(2)}{(4\pi \cdot 4/0.057)^2} = 1.0 \ \mu\text{W} = -30 \text{ dBm} \qquad (7.3)$$

Table 7.2   Details of WLAN Modulations and Some Required Receiver and Transmitter Characteristics

Modulation	Data Rate (Mb/s)	Receiver Adjacent Channel Rejection (dB)	Receiver Minimum Sensitivity (dBm)	EVM (dB)
BPSK	6.5	16	−82	−5
QPSK	13.0	13	−79	−10
QPSK	19.5	11	−77	−13
16-QAM	26.0	8	−74	−16
16-QAM	39.5	4	−70	−19
64-QAM	52.0	0	−66	−22
64-QAM	58.5	−1	−65	−25
64-QAM	65.0	−2	−64	−28

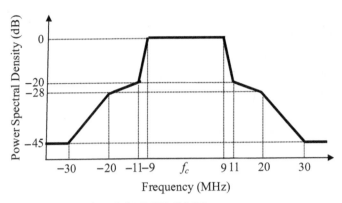

Figure 7.4 The transmit spectral mask for 5-GHz WLAN.

This also happens to be the maximum receive power of signals in undesired channels with which the radio must be able to deal, as specified in the standard. Now it will be possible to estimate the maximum distance that the link will function over. The simpler modulations will be used for greater distances to allow for some low data rate communication in favor of having the link fail altogether. For the simplest BPSK link the receiver must be able to detect a signal at –82 dBm. Assuming an indoor environment in which the path loss exponent ρ is 6, we can determine the maximum range for this link, noting that

$$P_{R\|dB} = P_{T\|dB} + G_{T\|dB} + G_{R\|dB} - L_{P\|dB}(d_0) - 10\rho\log_{10}\frac{d}{d_0} - L_{\sigma\|dB} \qquad (7.4)$$

Therefore,

$$10\rho\log_{10}\frac{d}{d_0} = P_{T\|dB} - P_{R\|dB} + G_{T\|dB} + G_{R\|dB} - L_{P\|dB}(d_0)$$

$$10\cdot 6\log_{10}\frac{d}{d_0} = 23\ \text{dBm} - (-82\ \text{dBm}) + 3\ \text{dBi} + 3\ \text{dBi} - 53\ \text{dB} = 58\ \text{dB} \qquad (7.5)$$

$$d = 37\text{m} = 121\ \text{feet}$$

Therefore a rough estimation for the range of this link would be 120 feet. The noise power at the receiver is –174 dBm/Hz before any corruption occurs in the receiver. In a 20-MHz channel, which has an effective noise bandwidth of 17.5 MHz, that is about –102 dBm. This would provide an SNR of 20 dB at the input of the receiver. At the maximum range, the radio is likely working with its simplest modulation (BPSK). BPSK is very robust and even with SNRs as low as 0 dB, it can still achieve a packet error rate of less than 10%. With some margin a good starting point might be to aim for a SNR of 10 dB at the output of the ADC. Therefore, NF for the entire radio must be 10 dB or better. Note here that we have not assumed any implementation loss in the baseband, but we will still use this number as a starting point.

A band-select filter will be attached to the antenna followed by a switch to select between receive and transmit mode. It can be expected that the switch, filter, and lines connecting the antenna to the input of the radio chip could have a loss of about 4 dB. Now let us specify the front end. The RF front end should have enough gain to overcome the high NF of parts that follow, but not too high to cause linearity problems for the baseband blocks, cause stability problems, or make the design of these amplifiers overly complicated. For this example, parts of the RF specifications will rather arbitrarily be set to be reasonable specifications for a front end in a modern CMOS process at 5 GHz. The overall specifications will then be achieved through adjustments in the baseband design. Let us specify the NF of the LNA as 2 dB. We will give it a gain of 18 dB and give it a low gain mode as well that has a gain of 2 dB to help when the signal level is high. Mixers have a relatively poor NF compared to LNAs and often lower gain. Let us set the NF of the mixer to be less than 7 dB and choose it to have a gain of 6 dB. From here, we will look at what the baseband components need to do and the performance that they will have to accommodate.

Now we need to determine what gain is actually required. We will assume that the ADC will require 0.5 Vrms at its input. Now –82 dBm in a 50Ω system means that the voltage level at the input will be 17.7 µVrms. Taking this to 0.5 Vrms requires a gain of $20\log(0.5/17.7\mu) = 89$ dB. With a net RF gain of 20 dB (18 dB of gain in the LNA plus 6 dB of gain in the mixer minus 4 dB of loss from the passives), the BB section needs a maximum gain of 69 dB. Now if the maximum receive power is –30 dBm, the gain must be able to be reduced by 52 dB. As there is 16 dB of range in the LNA, the gain of the baseband stage must be able to be reduced by 36 dB to 33 dB. We will assume that the gain can be adjusted in 3-dB steps. Therefore, the receiver gain range will be 37 to 89 dB.

Next we need to consider the noise figure. Doing a lineup on the NF at maximum gain (the place where the NF is the most important) yields the lineup shown in Table 7.3. Choosing the baseband filter to have a gain of 5 dB and a NF of 15 dB and choosing the baseband amplifiers to have an NF of 20 dB will yield an overall NF of 9.4 dB, thus maintaining a SNR of at least 10 dB at the input of the ADC.

Now let us consider the linearity requirement of the front end. Although larger in-band blockers are possible, according to the standard, our receiver only needs to tolerate blockers as large as –30 dBm. If we want the in-band blockers to provide an IM3 product of –92 dBm (10 dB below the minimum detectable signal level for an SNR of at least 10 dB) or lower, the IIP3 of the radio must be 1 dBm at the input. Assuming a 4-dB loss at the input before the LNA that puts the IIP3 of the LNA at –3 dBm, we will set it to be 0 dBm to avoid it being the limiting factor. The mixer will need to be 18 dB higher than this at 15 dBm. After the mixer, the input of the baseband will need to have an IIP3 of 21 dBm. This illustrates why we have to put

Table 7.3   Gain, Noise, and Linearity of the Receiver Lineup

	Band Select BPF	Switch	Lines to Radio Chip	LNA	Mixer	BB Filter	BB Amp	Total
Gain/Loss Max (dB)	–2	–1	–1	18	6	5	64	89
Gain/Loss Min (dB)	–2	–1	–1	2	6	5	28	37
NF Max Gain (dB)	2	1	1	2	6	15	20	9.4
NF Min Gain (dB)	2	1	1	8	6	15	35	26.5
IIP3 linearity due to blockers (dBm)	N/A	N/A	N/A	7	24	26	N/A	
IIP3 linearity due to large desired input signal (dBm)	N/A	N/A	N/A	–14	–12	–6	–1	

the filter in front of the gain stage; otherwise, the filter would have an unrealistic linearity requirement. Assuming that the baseband filter works properly, there is now no linearity requirement on the baseband amplifiers due to blockers. These specifications are also shown in Table 7.3.

Next consider the case of a large desired signal. Receiving a –30-dBm signal at the antenna means that we have –34 dBm of power at the LNA input. Thus, in the maximum signal mode, the 1-dB compression point should be –24 dBm assuming 10-dB backoff from the 1-dB compression point to get a good EVM with an OFDM signal. IIP3 will be 10 dB higher than this at –14 dBm. Thus, the mixer IIP3 will be –12 dBm (remember that the LNA will be in low gain mode), so it easily passes the linearity based on the blocker requirement, therefore so will the baseband. The baseband filter will have to have an IIP3 of –6 dBm and the baseband amplifiers will have to have an IIP3 of –1 dBm. These specifications are also shown in Table 7.3.

Now we need to look at the phase noise requirements of the LO. Now once more assuming that the adjacent channel is –30 dBm and the given channel is –82 dBm and needing a SNR of 10 dB, we get RLadj = –30 – (–82 – 10) = –62 dB. The adjacent channel will be 20 MHz wide so the maximum phase noise will be –62 – 10log(20 MHz) = –135 dBc/Hz. This is the phase noise at an offset of 10 MHz or higher. The spurs must be at –62 dBc or lower. Now in-band phase noise is due to SNR or EVM. An EVM of only 10 dB or 10% gives a phase noise of 0.1/10 MHz = 10 nrads/Hz or –80 dBc/Hz in band if the noise was flat until 10-MHz offset, which it will not be. This will be fine for low modulations, but for higher modulations it will not work well. Let us assume that we will want an EVM of 25 dB for stronger signals; then the in-band phase noise will be 0.00316/10 MHz = 316 prads/Hz or –95 dBc/Hz. We will assume a 40-MHz reference for the synthesizer, and assume that the in-band corner will be below 1 MHz. In this case, –90 dBc/Hz should be adequate. The out-of-band noise will be challenging if we assume –30 dBm blockers, but this number is not impossible. This also puts a limit of the phase matching of the LO of 0.0032 radian or 0.2°.

The LO can be simulated with the rest of the receiver in MATLAB. Modeling the performance of the LO using the theory presented in Chapter 3, the frequency-domain plots can be obtained to test that the realistic performance requirements are being modeled properly. The results are shown in Figure 7.5. Here Figure 7.5(a) shows a narrow bandwidth so that the in-band phase noise of –90 dBc/Hz can be clearly seen. The natural frequency of the synthesizer is set to 500 kHz and the LO tone power is set to 0 dBm. Note that the resolution of the frequency plot is 1 kHz. Figure 7.5(b) shows a wider bandwidth so that the reference tones at 40-MHz offset of –65 dBc are clearly visible. The phase noise floor of –135 dBc/Hz can also be seen. If the plot scale is changed, the frequency response of the LO synthesizer can be seen more clearly. This is shown in Figure 7.6.

The performance of the baseband filters and the ADCs also needs to be determined. If we have a –82-dBm signal with a –30-dBm blocker in the adjacent channel, we need to attenuate this signal before it gets aliased by the ADC. This means that this signal needs to be attenuated by at least –30 + 92 dBm = 62 dB to provide a SNR of at least 10 dB at the output of the ADC. Thus, we need an adjacent channel rejection of 62 dB. The baseband bandwidth is 10 MHz or more precisely 8.75 MHz to the edge of the passband, and the mixed down adjacent channel will begin at 11.25 MHz, which is the frequency the baseband filter will need to have 62 dB

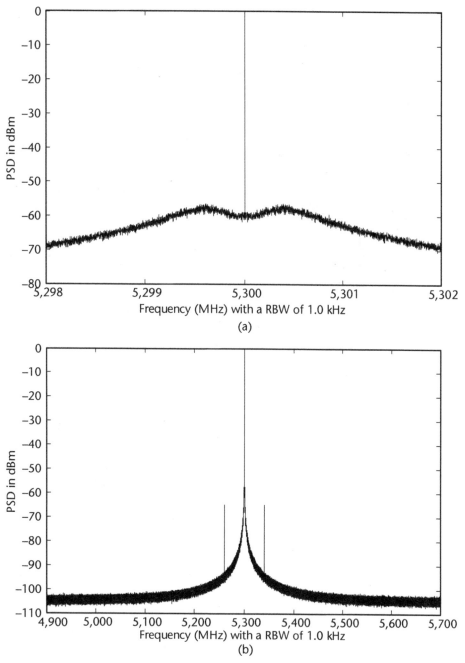

Figure 7.5 Frequency spectrum of the LO modeled in MATLAB with phase noise and reference spurs, shown with (a) a 4-MHz bandwidth and (b) a 600-MHz bandwidth.

of attenuation at. From simulation, a lowpass Butterworth filter with a passband edge of 8.3 MHz would require a thirtieth-order filter to meet this requirement. This is quite a high order. Simulating a Chebychev filter with 4-dB passband ripple means that only a tenth-order filter will be required. This will be our choice for the baseband filter. A plot of the frequency response of this filter is shown in Figure 7.7. We can also design the ADC assuming that we want at least 20-dB SNR. We will

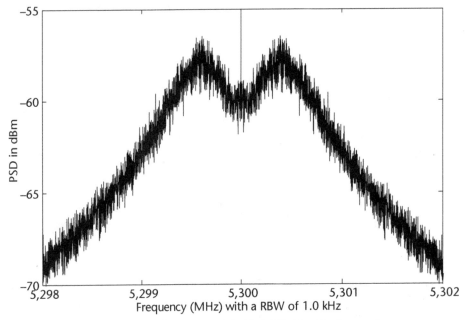

Figure 7.6 Frequency spectrum of the LO modeled in MATLAB with phase noise.

not oversample so the clock on the ADCs will be 20 MHz. A 4-bit ADC will give us a dynamic range of 6.02 * 4 + 1.76 = 25.8 dB, which should be acceptable.

Next the entire receive chain including the ADC was modeled in MATLAB to verify the design calculations. The first simulation was to apply a single CW tone with a power of –82 dBm to the input at 5,304 MHz with the LO set to 5,300 MHz

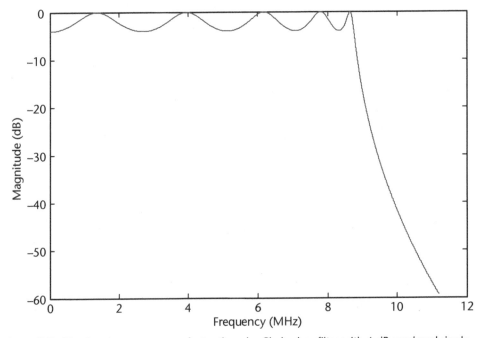

Figure 7.7 The frequency response of a tenth-order Chebyshev filter with 4-dB passband ripple.

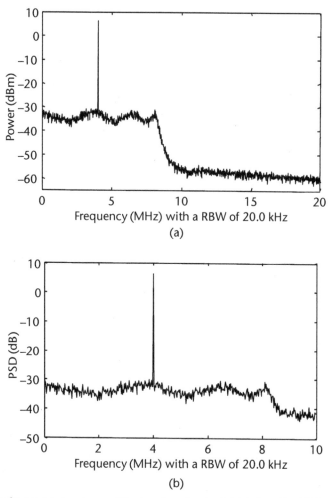

Figure 7.8 The frequency spectrum of the baseband output (a) before the ADC and (b) after the ADC.

and then compute the NF of the entire system. The frequency spectrum of the baseband output before and after the ADC is shown in Figure 7.8. In Figure 7.8(a) you can see the shape of the baseband filter. Measuring the decline in the SNR from the input to the output of the ADC confirms that the system has a NF of 9.6 dB and the overall gain is 88.1 dB. The input waveform to the ADC has an rms value of 0.5V, which is as designed.

Next the BB filter was tested. One signal was applied at 6-MHz offset and another one at 15.1 MHz offset from the LO. Noise was reduced so that simulation resolution was not an issue. Figure 7.9 shows that the –30-dBm signal that is out of band is still 15 dB lower than the in-band –82-dBm signal.

Next, two blockers were used in an attempt to jam the radio. Two tones were injected at 20-MHz and 35-MHz offset from the LO at a power level of –30 dBm. This will create an IM tone at 5 MHz. In addition a desired signal at 4-MHz offset from the LO at a –82-dBm power level was also injected into the radio. The results of this test are shown in Figure 7.10. Here we can see that the interfering tone is still

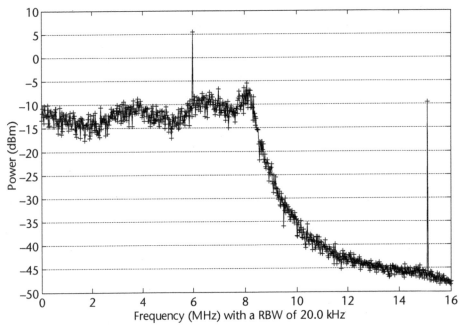

Figure 7.9 Plot showing a 15.1-MHz, −30-dBm signal being filtered and compared to an −82-dBm in-band signal.

11 dB below the desired tone. Note that this simulation was also run with reduced noise sources to avoid issues with resolution.

Next the radio was tested for desired signal linearity. In this test a signal was fed into the radio at different power levels. The output power before the ADC was measured to determine the 1-dB compression point of the radio. This was done in

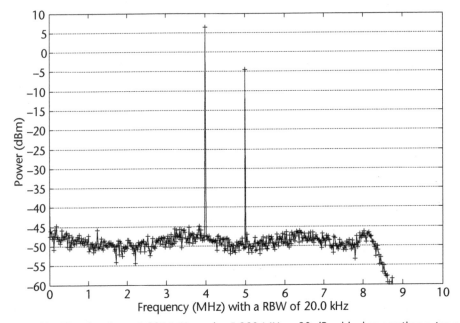

Figure 7.10 Plot showing a 5,335-MHz and a 5,320-MHz, −30 dBm blocker creating a tone at 5 MHz, compared to a 4-MHz, in-band signal at −82 dBm.

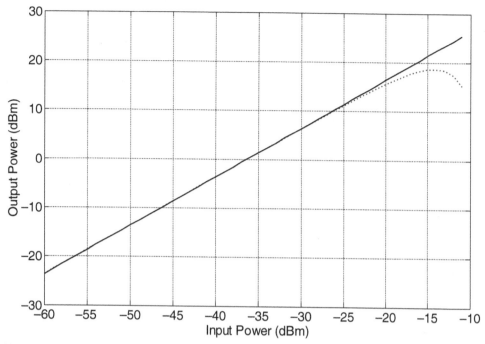

Figure 7.11 Plot showing output power at the input of the ADC versus input power. The plot is done at maximum gain and shows a 1-dB compression point of –19 dBm.

maximum gain mode. The results are shown in Figure 7.11. This plot shows that the radio achieves a 1-dB compression point of –19 dBm. This provides 11 dB of backoff at an input power of –30 dBm, which should be sufficient even for OFDM signals. The output of the ADC is also plotted for a sweep of input power and the result of this simulation is shown in Figure 7.12. Here because of the finite resolution of the ADC, small signals cannot be properly passed to the output. This reinforces one of the main reasons for a more complicated radio: to reduce the number of required bits in the ADC. Here the 1-dB compression level is –24 dBm, which is slightly worse than the radio chain simulated without the ADC.

Next a tone was fed into the radio at –30 dBm, 6 MHz away from an LO spur at 40-MHz offset from the LO itself. This was compared to the amplitude of a desired tone at –82 dBm, 4 MHz away from the LO itself. The results of this test are shown in Figure 7.13. This verifies the calculations for the LO spur levels as being adequate to make sure that a blocker would still be at least 10 dB below the desired input with this spur level.

Next the transmitter will be considered. First, we will start by assuming a 4-bit DAC as in the receiver. Assuming a 0.5 Vrms output from the DAC (7 dBm), to transmit 23 dBm means that our chain will need 16 dB of gain. We will reuse the filters designed for the receiver and therefore they will provide 5 dB of gain. If we then budget 6 dB of gain for the mixers and 10 dB of gain for the PA, then we will need a loss of 5 dB in the baseband to achieve an output power of 23 dBm. This PA will need an output 1-dB compression point of about 33 dBm (10-dB backoff) because it will need to be linear and not corrupt the EVM of the transmitted waveform. Thus OIP3 will be roughly 43 dBm. The ACPR with this linearity will be

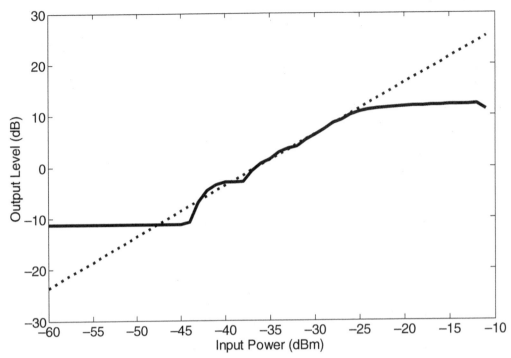

Figure 7.12 Plot showing the signal level at the output of the ADC versus input power. The plot is done at maximum gain and shows a 1-dB compression point of –24 dBm.

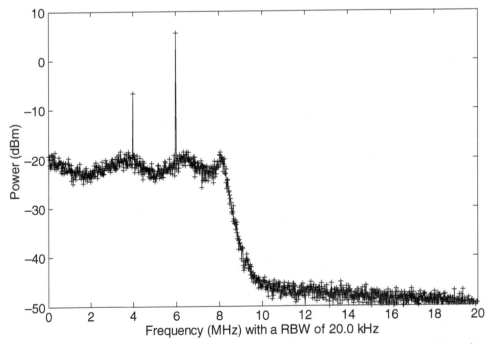

Figure 7.13 Plot showing desired –82-dBm tone input at 5,206 MHz with interferer injected at 5,156 MHz (4 MHz from LO spur at –65 dBc).

23/2 – 2 • (43) = –74.5 dB. This should be more than sufficient to meet the spectral mask requirements of –40 dB, so this should be fine. We will put a –5-dB variable gain attenuator in the baseband to make up the difference. This attenuator should be made programmable with a variable level of gain/loss to overcome any process or temperature variations. The lineup is shown in Table 7.4. Note that the LO requirements have already been discussed as the receiver was being designed.

Now that the paper design is complete, some simulations can be run to verify the functionality of the radio. To do this, an OFDM waveform needs to be created. To test the system, 16-QAM modulation will be used. Random bits will be generated and then modulated to produce I and Q components. Once this is done, 56 symbols will be used in conjunction with a zero at DC and seven other zeros will be put through an inverse FFT function. This must be done for all the data to produce the time-domain OFDM symbols. The symbols will be sampled at a rate of 50 ns. This will make the frequency spectrum of the waveform correct. This waveform will then be passed through an interpolation filter and upsampled to produce a time-domain waveform with a bandwidth less than 10 MHz and no ISI.

After this has been completed, the signal is then passed through the baseband filter and then amplified and upconverted to RF using a LO with phase noise and applied to the power amplifier. Once this done to test the RF waveform, we can plot the frequency response. It is shown in Figure 7.14, which shows that the spectrum will pass under the spectral mask so the ACPR is fine.

To test the EVM of the transmitter, an ideal receiver is added to the simulation. The waveform is mixed back down to baseband. Once there, the signal is sampled and then passed through a 64-point FFT. The sampling point in the simple simulation is set manually to align the received signal with the transmitted one. The received signal, transmitted signal, and ideal source pulses are all shown in Figure 7.15. Note that after alignment the waveforms are very close, but not perfect due to the nonideal properties of the RF transmitter. The IQ plot is shown in Figure 7.16. It shows that the received signal has an EVM of –24.2 dB.

7.2 A Basic GPS Receiver Design

This section discusses some issues related to the design of GPS receivers.

7.2.1 GPS Overview

Navstar Global Position System, referred to as GPS, is owned by the U.S. government and operated by the U.S. Air Force. It was first launched in 1978. Two GPS services are provided: Precise Position Service (PPS), which is available to the military, and Standard Position Service (SPS) for civilians, which was designed initially

Table 7.4 Gain, Noise, and Linearity of the Receiver Lineup

	BB Attenuator	BB Filter	Mixer	PA and RF Gain Stages	Total Gain
Gain/Loss typ (dB)	–5	5	6	10	16
Linearity (Output P1dB) (dBm)	12	17	23	33	N/A

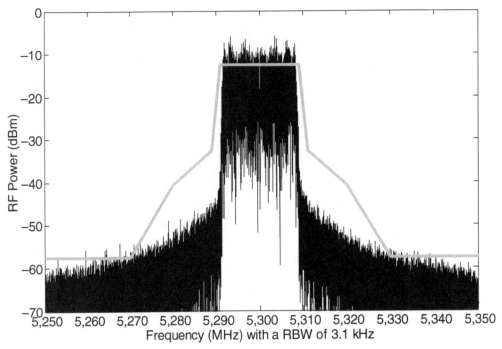

Figure 7.14 Transmitted RF spectrum with 23-dBm output power.

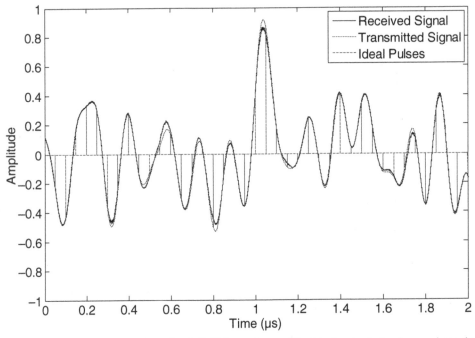

Figure 7.15 Transmitted and received OFDM waveforms. The ideal pulse train shows where the signal is sampled.

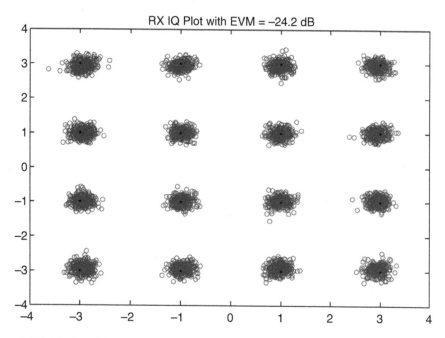

Figure 7.16   IQ plot of the received 16QAM signals.

to be less accurate through Selective Availability, which was discontinued in 2007. SIS is Signal In Space and PS is Performance Standard.

The C/A (coarse acquisition) code is binary phase shift keying direct sequence spread spectrum (BPSK DSSS) at 1.023 MHz resulting in a main lobe about 2 MHz wide. There is also a P (precision) code, spread at 10.23 MHz for a 20-MHz bandwidth, but that is not available to the general public. In the following, the discussion will be limited to the C/A code.

A GPS signal is 50 b/s of navigation data, spread by a spreading code of 1.023 Mb/s as shown in the figure. The spreading code is a maximum length sequence or Gold code generated in software, or in hardware with a feedback shift register. Taps can be selected to generate different codes, of which one of 32 will identify each satellite. The spread data is BPSK modulated by multiplying the spread data with the 1.57542 GHz carrier, which is then transmitted. This GPS band is referred to as the L1 band. Note that the L2 GPS band at 1.2276 GHz will not be discussed here. Figure 7.17 shows the transmitter and Figure 7.18 shows navigation data being spread by the spreading code.

In the receiver, shown in Figure 7.19, the antenna picks up signals from multiple satellites. The signal is amplified, downconverted, and digitized. The BPSK signal is

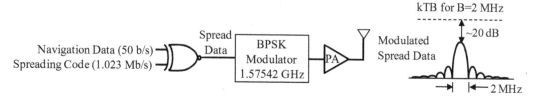

Figure 7.17   Simplified GPS Signal Generation in the L1 band at 1.57542 GHz.

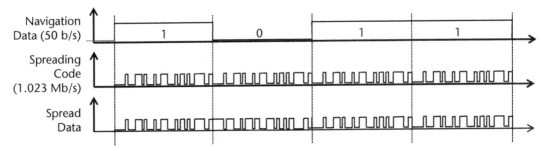

Figure 7.18 Generation of GPS signal. Note that, for clarity, codes are not shown to scale.

demodulated and then the signal is correlated with local copies of the different possible spreading codes to receive the navigation data from various satellites. To allow determination of position and time, at least four sets of navigation data are needed from at least four satellites, but at any time, up to 10 or 11 may be received. Note that the codes are designed so that different codes have very low correlation and the codes have very low auto-correlation; thus, correlation will be strong only when the local code is aligned with the incoming sequence. The 1.023-Mb/s code is repeated every millisecond. During acquisition, if the local code is not correlated with the incoming signal, the local code is time shifted by 1 bit and another correlation is done. This time shifting is done until a strong correlation is seen at which point the modulo two addition (XOR) will result in the 50-b/s navigation data. This data is then further processed to generate information about location, altitude, velocity, and time.

7.2.2 RF Specification Calculations

Thermal noise floor is calculated as

$$\text{Noise Floor} = -174 \text{ dBm} + 10\log_{10}(2 \text{ MHz}) = -111 \text{ dBm} \qquad (7.6)$$

Front-end noise figure is added to this to affect the minimum signal that can be received. The typical numbers quoted are a few decibels up to more than 5 dB.

The receiver is required to provide roughly a full scale voltage to the DAC.

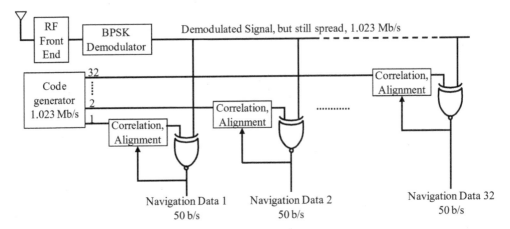

Figure 7.19 Simplified block diagram of GPS receiver.

$$G = P_{DAC} - P_{IN} \tag{7.7}$$

Before correlation, noise is dominant so both signal and noise needs to be converted. The dominant input Pin is given by thermal noise kTB. This noise floor is in the microvolt range and, assuming the required ADC has a full scale input of 400 mV, a total cascaded gain of about 103 dB is needed. AGC is needed to account for gain variations from the antenna, temperature, fabrication process, power supply, and ADC input range. Over the range of −45° to +85° the input power changes by only about 2 dB. Otherwise, the only variation is due to process voltage and temperature of the actual electronics; in the absence of jammers or interfering signals in band or close by, there is not a lot of gain variation, but a margin must be left so 10 dB should be adequate. If there are jammers, obviously more would be required.

Often only a 2-bit ADC is adopted, although for higher precision, 3 or 4 bits are necessary to minimize CNR loss in the ADC.

Architecture Selection

As will be seen, the required specifications in most cases are not that stringent allowing a fully or nearly fully integrated solution. If the IF is chosen to be at 100 to 200 MHz, then the image is sufficiently far away that a low Q-filter can be used in front of the LNA. This would mean there needs to be a second stage of downconversion. At the other extreme, direct downconversion is typically not used because most of the energy is close to the carrier, resulting in high $1/f$ noise. An intermediate solution is to choose a low IF architecture. Depending on the choice of the low IF, this may require image reject mixers or complex filters. Because the overall GPS function is in a 20-MHz band, any IF chosen as 10 MHz or less will result in the image being in the GPS band at a known (low) level. As well, filters at such low frequencies are relatively easy to construct and can have low power.

By itself, a GPS receiver can be straight forward in the sense that there is only one channel and therefore there is no intermodulation. However, when combined with a cellular band transceiver, there are interfering GSM and PCS cellular bands in the 1.9-GHz range at an offset of roughly 330 MHz from the GPS carrier setting a minimum phase noise for the LO to avoid reciprocal mixing. With an off-chip front-end filter, these blockers can be reduced to the point where they are not significant, but such rejection can be quite difficult in a fully integrated solution, and compounded if the GPS receiver is part of a transceiver (e.g., cell phone) also transmitting at one of these blocker frequencies. Zero IF receivers are typically not used because most of the energy in the GPS signal is right at the carrier, and because direct downconversion would result in a lot of $1/f$ noise. More typically, low IF architectures are used. In recent publications, intermediate frequencies were chosen in the 1- to 10-MHz range so that the component at 0 Hz could be removed and to allow a fairly low sampling frequency f_S in front of the A/D converter. Of course, such receivers need image rejection, but since there are no other channels or interferers at this frequency, the image is typically only the noise from within the GPS channel itself (meaning that an image rejection of 20 dB would be plenty). The solution typically relies on some filtering at RF plus a complex baseband filter to reduce signals both in the image frequency band at $-f_{IF}$ and in the aliasing band at

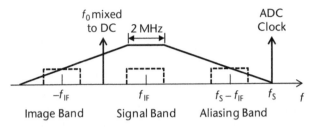

Figure 7.20 Filtering requirements at baseband.

f_S-f_{IF}. When possible, a small multiple of the 1.023-MHz chip-rate signal is used, for example, a multiple of four results in 4.092 MHz whereas a multiple of 10 results in 10.23 MHz. Filtering requirements at baseband are shown in Figure 7.20. If the IF is at 10 MHz or less, then the aliasing and image band are also within the overall 20-MHz GPS signal band, hence the interference should be at the noise level so filtering should be relatively straightforward, and a reduction by a factor of 10 (20 dB) should be adequate.

The received signal power is typically –130 dBm while the noise in the 2-MHz band is kTB is –111 dBm; thus, the signal is 19 dB below the noise. However, the signal is so low because the 50 bits per second are spread by the 1.023 MHz C/A code for a processing gain of 1.023M/50 = 20,460 (taking 10 log results in 43.11 dB, making SNR ideally + 24.11 dB). Thus, the effective signal is at –130 dBm + 43.11 dB = –86.89 dBm, about 24 dB above the noise level.

Synthesizer Design

For the BPSK modulated signal on a 1.575-GHz carrier, it is straightforward to design the synthesizer at the double frequency and then to divide by 2 to provide four phases. For BPSK, quadrature phase error is not extremely critical with recent publications quoting phase errors of less than 5°. For low IF conversion, in-band phase noise converts the noise level from other frequencies into the IF band as illustrated in Figure 7.21. Because the IF band of 2 MHz adds another 63 dB of noise, if the in-band phase noise were at –63 dBc/Hz, the noise would be doubled. To reduce such added noise, typically a further factor of the order of 20 dB is used, resulting in an

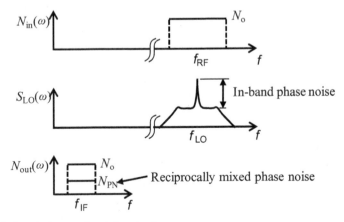

Figure 7.21 In-band phase noise determination.

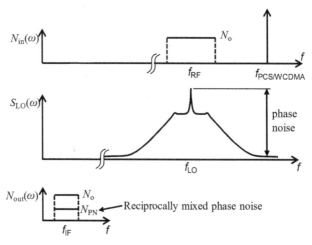

Figure 7.22   Phase noise requirements for cellular band blockers at 330-MHz offset.

in-band phase noise specification of about –85 dBc/Hz both for in-band as well as the specification for an offset of 1 MHz. This is a very straightforward design and, as a result, recent papers have shown in-band phase noise of at least –90 dBc/Hz and more typically –105 dBc/Hz or better at a 1-MHz offset.

Phase noise at larger offsets may have to be considerably better if large interfering signals in the GSM and PCS cellular bands are present at a typical offset around 330 MHz as illustrated in Figure 7.22. If the GPS receiver is part of a transceiver that also generates these cellular signals, the levels can be at broadcast at up to 30 dBm; however, assuming there is a front-end filter, a more likely internal signal at the GPS LNA input would be about –10 dBm. The LNA is likely designed as a tuned circuit and with reasonable on-chip components could provide another 20 dB or so filtering at the 330-MHz offset resulting in a blocker level of –30 dBm as shown in Figure 7.6. The integrated reciprocally mixed noise is required to be significantly less than the original kTB noise at –111 dBm. Integrated across the 2-MHz bandwidth adds another 63 dB, so for the reciprocally mixed phase noise to be equal to the kTB noise at –111 dBm would require a phase noise of (–111+30 – 63) dBm or –144 dBm. Thus phase noise of –150 dBc/Hz at a 330-MHz offset would leave a margin of 6 dB. This can be improved by further reducing phase noise, or increasing the attenuation of the LNA to more than 20 dB.

References

[1]   IEEE Std. 802.11-2007, *IEEE Standard for Information Technology—Telecommunications and Information Exchange Between Systems—Local and Metropolitan Area Networks—Specific Requirements. Part 11: Wireless LAN Medium Access Control (MAC) and Physical Layer (PHY) Specifications*, June 2007.

[2]   IEEE Std. 802.11n-2009, *IEEE Standard for Information Technology—Telecommunications and Information Exchange Between Systems—Local and Metropolitan Area Networks—Specific Requirements. Part 11: Wireless LAN Medium Access Control (MAC) and Physical Layer (PHY) Specifications. Amendment 5: Enhancements for Higher Throughput*, September 2009.

[3]   http://www.gps.gov/technical/ps/

About the Authors

John W. M. Rogers received a Ph.D. in 2002 in electrical engineering from Carleton University, Ottawa, Canada. Since 2002, he has been a member of the faculty of engineering at Carleton University where he is now an associate professor. He is the coauthor of *Radio Frequency Integrated Circuit Design, Second Edition* (Artech House, 2010) and *Integrated Circuit Design for High-Speed Frequency Synthesis* (Artech House, 2006). His research interests are in the areas of RFIC and mixed-signal design for wireless applications.

Calvin Plett received a B.A.Sc. in electrical engineering from the University of Waterloo, Canada, in 1982, and an M.Eng. and a Ph.D. from Carleton University, Ottawa, Canada, in 1986 and 1991, respectively. In 1989 he joined the Department of Electronics, Carleton University, Ottawa, Canada, where he is now a professor and the chair of the department. He has also spent many years doing consulting work and collaborative research in RFIC design with numerous students and an array of companies. He is a coauthor of the books *Radio Frequency Integrated Circuit Design* (Artech House, 2003), *Radio Frequency Integrated Circuit Design, Second Edition* (Artech House, 2010), and *Integrated Circuit Design for High-Speed Frequency Synthesis* (Artech House, 2006). His research interests include the design of analog and radio-frequency integrated circuits, including filter design, and communications applications. Dr. Plett is a member of the AES and the PEO and a senior member of the IEEE.

Ian D. Marsland received a B.Sc.Eng. (Honours) in mathematics and engineering from Queen's University in Kingston in 1987. From 1987 to 1990, he was with Myrias Research Corporation, Edmonton, and CDP Communications, Toronto, where he worked as a software engineer. He received an M.Sc. in 1994 and a Ph.D. in 1999, both in electrical engineering from the University of British Columbia. He joined the Department of Systems and Computer Engineering at Carleton University in 1999, where he is now an associate professor. His research interests are in the area of wireless digital communication.

Index

Solid-State Microwave High-Power Amplifiers, Franco Sechi and Marina Bujatti

Stability Analysis of Nonlinear Microwave Circuits, Almudena Suárez and
  Raymond Quéré

Substrate Noise Coupling in Analog/RF Circuits, Stephane Bronckers,
  Geert Van der Plas, Gerd Vandersteen, and Yves Rolain

System-in-Package RF Design and Applications, Michael P. Gaynor

TRAVIS 2.0: Transmission Line Visualization Software and User's Guide, Version 2.0,
  Robert G. Kaires and Barton T. Hickman

Understanding Microwave Heating Cavities, Tse V. Chow Ting Chan and
  Howard C. Reader

For further information on these and other Artech House titles, including previously
considered out-of-print books now available through our In-Print-Forever® (IPF®)
program, contact:

Artech House Publishers	Artech House Books
685 Canton Street	16 Sussex Street
Norwood, MA 02062	London SW1V 4RW UK
Phone: 781-769-9750	Phone: +44 (0)20 7596 8750
Fax: 781-769-6334	Fax: +44 (0)20 7630 0166
e-mail: artech@artechhouse.com	e-mail: artech-uk@artechhouse.com

Find us on the World Wide Web at: www.artechhouse.com